Wildlife Necropsy and Forensics

This book is the result of simultaneous compilation of information on the postmortem examination of wild animals for over 20 years ever since the author attended the summer institute on the pathology of diseases of wildlife at the College of Veterinary Science, Hebbal, Bangalore during May 1979 and also the experience gained while conducting the postmortem examination of wild animals submitted from the S.V.Zoo park at Tirupati. The book is written in 2 parts. In Part I the necropsy procedures in some wild animals along with the procedure for collection and preservation of the collected specimens for diagnostic purposes are given. These procedures are common both to the disease investigation as well as forensic necropsies in wild animals except that the forensic necropsies need more careful examination and collection and preservation of material for diagnostic purposes. In the second part information on the list of various wildlife crimes, the role of each member of wildlife crime team investigation in general and the role of pathologist in particular, the forensic necropsy and collection of evidential items and the methodology for crime investigation is given.

Dr. P. K. Sriraman is a retired (Voluntary) Associate professor and Head, Department of Pathology, College of Veterinary Science, Rajendranagar, Hyderabad-30 after putting in 29 years of active service involving teaching (both to the undergraduate and post graduates), research and extension activities.

Wildlife Necropsy and Forensics

P.K.Sriraman
Associate Professor and Head (Retd.)
Department of Pathology
College of Veterinary Science
HYDERABAD - 500030
Telengana (India)

CRC Press is an imprint of the
Taylor & Francis Group, an **informa** business

NARENDRA PUBLISHING HOUSE

First published 2021
by CRC Press
2 Park Square, Milton Park, Abingdon, Oxon, OX14 4RN

and by CRC Press
6000 Broken Sound Parkway NW, Suite 300, Boca Raton, FL 33487-2742

© 2021 Narendra Publishing House

CRC Press is an imprint of Informa UK Limited

The right of P.K. Sriraman to be identified as author of this work has been asserted by him in accordance with sections 77 and 78 of the Copyright, Designs and Patents Act 1988.

Reasonable efforts have been made to publish reliable data and information, but the author and publisher cannot assume responsibility for the validity of all materials or the consequences of their use. The authors and publishers have attempted to trace the copyright holders of all material reproduced in this publication and apologize to copyright holders if permission to publish in this form has not been obtained. If any copyright material has not been acknowledged please write and let us know so we may rectify in any future reprint.

All rights reserved. No part of this book may be reprinted or reproduced or utilised in any form or by any electronic, mechanical, or other means, now known or hereafter invented, including photocopying and recording, or in any information storage or retrieval system, without permission in writing from the publishers.

For permission to photocopy or use material electronically from this work, access www.copyright.com or contact the Copyright Clearance Center, Inc. (CCC), 222 Rosewood Drive, Danvers, MA 01923, 978-750-8400. For works that are not available on CCC please contact mpkbookspermissions@tandf.co.uk

Trademark notice: Product or corporate names may be trademarks or registered trademarks, and are used only for identification and explanation without intent to infringe.

Print edition not for sale in South Asia (India, Sri Lanka, Nepal, Bangladesh, Pakistan or Bhutan).

British Library Cataloguing-in-Publication Data
A catalogue record for this book is available from the British Library

Library of Congress Cataloging-in-Publication Data
A catalog record has been requested

ISBN: 978-0-367-77596-4 (hbk)
ISBN: 978-1-003-17201-7 (ebk)

Contents

	Preface	*iiv*
	Part I Wildlife Necropsy	**1 - 100**
1	Introduction	3
2	Collection and Preservation of Biological Specimens	24
3	Necropsy Methods	50
4	Autopsy Report	97
	Part II Wildlife Forensics	**101 - 305**
1	Introduction	103
2	Crimes Against Wildlife	105
3	Forensic Analysis of Wildlife Crimes	116
	1. Role of Pathologist	119
	2. Role of Morphologist	126
	3. Role of Molecular Biologist (Geneticist)	130
	4. Role of a Criminalist	135
	5. Role of an Analytical Chemist	142
	6. Role of the Forensic Specialist / Investigator	147
4	Forensic Necropsy	152
5	Examination of the Wounds	200
6	Types of Firearms and Terminology	241
7	Poisoning	258
8	Record Keeping	269
9	Wildlife Crime Investigation	273

Appendices	**306 - 364**
Appendix I	306
Appendix II	314
Appendix III	318
Appendix IV	342
Appendix V	352
Appendix VI	353
Appendix VII	354
Appendix VIII	359
Appendix IX	362
Appendix X	363

Preface

Necropsy (Postmortem examination) of wild animals for legal purposes has become increasingly common. Special procedures are necessary during such necropsies to ensure that the information collected is suitable for use in a court of law. Forensic necropsies should be performed by a trained pathologist who has experience with the species being examined since his credentials will be examined if a case reaches the court of law. Determination of the cause of death in wild animals is often difficult and may require the close co-operation of a number of disciplines. For the diagnostic laboratory to contribute fully to final diagnosis, the specimens collected must be selected carefully and preserved in suitable conditions.

Crimes involving wild animals, including the trade in animal parts like gallbladder of bears and paws, and poaching of protected and endangered species have increased in the recent past. There is also increasing interest in establishing the culpabilities for the death of wild animals as a result of pollution or poisons. These developments have resulted in greater need for evidence obtained by the examination of dead wildlife that will be acceptable in the court of law.

The book is written in 2 parts. In Part I the necropsy procedures in some wild animals along with the procedure for collection and preservation of specimens for diagnostic purposes are given. These procedures are common both to the disease investigation as well as forensic necropsies in wild animals except that the forensic necropsies need more careful examination and collection and preservation of material for diagnostic purposes. In the second part information on the list of various wildlife crimes, the role of each member of wildlife crime team investigation in general and the role of pathologist in particular, (the forensic veterinary pathology)and the salient features of methodology of crime investigation is given. Information on the mistakes and omissions in forensic necropsy, negative necropsy, common errors committed by pathologist while performing forensic necropsy, postmortem appearances of common pathological conditions and estimation of age of some of the domestic animals and their foetuses is given for guidance. Various types of wounds, and the types of firearms and their terminology are also given for easy understanding about the wounds and fire arms.

Hope this book will be of use to all those interested in wildlife preservation and especially to veterinary pathologists who are involved in conducting necropsies pertaining to wildlife crimes, wildlife personnel dealing with the law enforcement, students and teachers in veterinary colleges as well as the students of wildlife school and zoo personnel.

Dr. P.K. Sriraman

About the Author

Dr. P. K. Sriraman is a retired (Voluntary) Associate professor and Head, Department of Pathology, College of Veterinary Science, Rajendranagar, Hyderabad-30 after putting in 29 years of active service involving teaching (both to the undergraduate and post graduates), research and extension activities. The author was direct disciple of Professor Ganti A Sastry, a doyen in the field of veterinary pathology and Dr P. Rama Rao, and also worked with them till their retirement. The author was a recipient of Meritorious research worker award of ANGRA University, Hyderabad (1993), Gold medal for securing highest OGPA in PhD, ICMR Senior fellowship during PhD and "ICAR Merit scholarship during graduate studies. He has guided 27 MVScs and 5 PhD students during his tenure in the Veterinary colleges and published more than 120 research papers in journals of repute. He has authored 5 books. 1. "A guide to postmortem examination of animals" along with a CD ROM containing procedures for Postmortem examination of animals (first of its kind) 2 . "Veterinary laboratory diagnosis". Published by Jaypee Brothers Medical publishers (P) Ltd, New Delhi. 3. Lecture notes on Veterinary Pathology , 4. Lecture notes on Veterinary special pathology and 5.Diseases of wild animals and birds (Jaya Publishing House, Delhi).

Part I

Wildlife Necropsy

1 | Introduction

The disease affects the viability of wild animal population. In a balanced ecosystem most populations survive with low levels of disease or with periodic epidemics. With habitat restrictions, the wildlife populations have become more dense leading to an increased risk of catastrophic epidemics. With restricted habitat of wildlife there are increased chances of disease transmission between wild and domestic animals. To determine the disease risks to a population, one must know (1) the causes of morbidity and mortalities in that population and (2) the natural history of infectious diseases in that environment, including the history of previous epidemics. Most often it is necessary to examine the entire animal to know the cause of epidemic.

Determination of cause of death of a wild animal is often difficult and need cooperation of a number of disciplines. A careful selection and preservation of specimens in suitable condition sent to the laboratory will help in the final diagnosis of cause of death. Many a times the carcass of a wild animal is found in an unsuitable condition for postmortem examination due to high ambient temperatures which cause a very rapid putrefaction of carcasses, common in tropics and due to the presence of avian and mammalian scavengers. The carcasses of young animals disappear without a trace. In view of this, every effort should be made to examine even a single carcass. Efforts should be made to detect and destroy a representative sick animal for postmortem examination (PME). The mortality rate and structure of population particularly of young, pregnant/lactating female and old animals can be altered due to extreme cold climate, wet and draught and nutritional stress due to deep snow cover or draught which hinders the growth of vegetation. Hence attention should be paid to local environmental factors. Similarly, epizootic disease, caused by an agent which is enzootic in contiguous domestic stock, may act similarly.

The determination of cause of death in non-domestic animals in captivity or natural habitat is difficult due to (1)Lack of history of the disease (2) Lack of observation before death, (3) Non exhibition of symptoms of illness until they are weak to walk, fly or crawl (4) Sudden death without showing gross and microscopic lesions and (5) Non familiarity of normal anatomy of wild animals by examiners.

Many disease epidemics affecting the wildlife or livestock go undetected if appropriate samples are not collected for diagnostic testing from animals that died during epidemics. Selective sampling limits the information that could be procured from a wild animal necropsy which could aid in future population or ecosystem management. In the case of captive animals, the veterinarian will focus on that one area (eye, brain, heart, etc)where the source of the problem originated. But it is necessary to examine the entire animal. In the case of a wild animal no idea of cause of death may be known, while in captive animals the apparent cause of illness during life may simply mask the true cause of death. With appropriate samples and accurate written and photographic records the cause of an epidemic can be determined in most of the cases.

Wild animals rarely exhibit the signs of disease. Death may be as a result of an encounter with a predator that has taken advantage of the weakened state of the sick animal. In those species of animal with high reproductive rate, there is usually higher mortality among young ones. The carcasses of small and young animals disappear without traces.

In the investigation of any disease outbreak in wildlife, attention should therefore be paid to the local environmental circumstances, including those that may be affected by climate and man. The investigator must collect as much general and local information as possible, including any evidence of similar mortality or morbidity in domestic stock. When there is mass mortality it is important to examine (1). fresh carcasses of a number of animals, representative of the range of species affected and (2). the ages of individuals affected. One should remember that more than one disease process may be acting at any one time and that the major cause(s) of death may change during a prolonged mortality. The results of the postmortem examination with the history of the animal and assessment of the environment will help in determining their significance and recommending the future course of action. A thorough inspection of the surroundings of a sick or dead animal is vital. One must observe for signs of a struggle which may indicate an encounter with a predator or a paroxysm at the time of death. An accumulation of faeces behind the carcass may indicate a period of immobility, as may heavy browsing of food plants within easy reach. If the animal is found alive, it should be observed from a distance for any peculiarities of gait, respiration, excretion, unusual behaviour and unusual degrees of wildness or tameness. Photographs should be taken, illustrating environmental conditions and the appearance of sick animals and their lesions since these will be of great value if litigation is likely.

The performance of a thorough necropsy (post-mortem examination) plays an important part in the diagnosis of the cause of death. However, it is very important to avoid preconceptions that may limit the information and specimen collection. Determination of the cause of death in zoo animals is often difficult and may require the close co-operation of a number of disciplines. Some zoo veterinarians perform a *post-mortem* examination (PME) in the zoo. But in doing the *post-mortem*, careless observation and sampling can result in useless, and sometimes even harmful, information. Those who perform post-mortem examinations in this manner are running a risk, as overlooking fundamental changes in tissue can be costly. The specimen(s) collected must be selected carefully and preserved in suitable conditions. A thorough *post-mortem* examination of animals that die or are euthanized is a necessary adjunct to any good clinical practice. When a *post-mortem* is performed in the zoo itself, or samples are collected for diagnostic purposes, the results and information deriving from this activity are highly protocol sensitive. For a diagnostic laboratory to contribute fully to the final diagnosis, the specimen(s) collected must be selected carefully and preserved in suitable conditions. Collect complete tissue and blood samples from carcasses. If only selected samples are taken because a certain disease is suspected and if the animal does not have that disease, the specimens submitted to the laboratory may not be adequate for the diagnosis of other unsuspected diseases that may be involved in an epizootic. The pathologist uses the clinical history (including haematology, blood chemistry and therapeutic measurements), the gross description, culture results and other data, as well as the cytological and histological appearance of the lesions, to make a diagnosis.

Absence of any of these, or incorrect submission of tissues, will hamper this process.

Postmortem examination may be performed by the case clinician or by a specialist/ pathologist. When the gross postmortem is performed by an individual other than the pathologist (who will perform further examinations on samples provided), communication between the two parties is essential to ensure that optimal samples are taken (tissue type, volume/ weight, storage, transport, temperature etc.). One cannot reach accurate diagnosis, if an inadequate or incorrect samples are collected.

Reasons for Carrying Out Post-mortem Examination(PME)

- To know the cause of death. It can reveal genetic, environmental and dietary problems which can be avoided or controlled in future generations.

- Without a postmortem one may not know what has happened to the dead animal. For the protection of the species it is necessary to know if local villagers or poachers are placing poisoned carcasses out (for tigers). Poisoning of meat and waterholes is a common poaching method.

- To investigate an unsuccessful treatment. A post-mortem can reveal the success or failure of certain management practices. Without this type of examination, it's impossible to know for sure if the treatments which were chosen for an animal were a success or not.

- It helps in detecting the subclinical disease and in confirming the diagnosis. Much of the current available knowledge on medications, diets and surgeries comes as a result of what veterinarians have found during post-mortem examinations. Armed with this knowledge one may modify the techniques for the future.

- Zoonotic diseases like rabies can be tested for the presence. Rabies is detected quickly only by examining the brain during a post-mortem examination.

- It is done to confirm a diagnosis. A post-mortem may reveal a disease that has the potential to kill an entire wild animal group. Once the source of a problem is identified veterinarians in captive facilities can use vaccination to halt the illness before it takes hold. In zoos the death of one animal from an unidentified illness could mean a threat exists which would spread to all the animals of that species in the facility. In the wild it can be important to identify a possible disease which might eradicate an isolated animal population (of that species). Even the

Introduction   7

death of one animal from that species significantly decimates the available gene pool of animals.

- Post mortem examination is particularly important for animals which die in quarantine (in preparation for introduction to a collection, translocation or reintroduction program).

- By doing post-mortem examination, one can increase the knowledge about the disease.

Criteria

A necropsy should be performed, at the discretion of the attending veterinarian in the following circumstances.

- When a significant number of unexplained deaths is occurring.

- When there is a high death loss.

- When a strong chance exists that an undiagnosed infectious disease is present (with or without potential zoonoses).

- When the circumstances around a death indicate that a violation may have contributed to the death.

- When warranted by circumstances it may be performed by the case clinician or by a specialist/ pathologist.

Before actually conducting the postmortem examination on the body of the animal one should check whether there is a requisition to do so from proper authority. This is very important. In case of insured animal, a letter of request from the competent authority of the concerned insurance company should be obtained. In private case of uninsured animal the veterinarian should get permission from the controlling authority. If one is unable to attend the postmortem due to unavoidable circumstances, ask the police / aggrieved person to contact some other veterinary doctor.

Questions to be asked Before Post-mortem examination commences.

(1) Do you have permission.
(2) Do you know enough about the species to be confident to perform necropsy and
(3) Are you clear as to what questions you will be asked. If any doubts, delay the PME.

Points to Remember

1. Post-mortem examination should be conducted in good daylight / a broad daylight. If possible, findings should be dictated to an assistant during the examination, or (if not possible), noted down at the time of the examination. Usually the postmortem examination is not done when the light is dim since the colour changes cannot be appreciated. In case the veterinarian feels that the sun may set before the completion of postmortem examination, the veterinarian may return the requisition letter by mentioning the time and date and with remarks that "the postmortem examination would be possible the next day at certain time (mention time) since continuation of the postmortem in the artificial light during the night may be difficult since colour changes in the organs cannot be appreciated". Always record the time and date of death of the animal.

2. In case of poisoning, collect circumstantial evidences. It is the duty of the constable at the spot to identify the animal and ownership of the carcass.

3. It is strongly recommended but not required that a necropsy be performed on all elephants.

4. Forensic post-mortems (for legal investigation) should be performed by an experienced wildlife pathologist, since the credentials of the pathologist will be assessed as part of the case in the court. Similar protocols should be used for the post-mortem examination of domestic, free-ranging or captive wild mammals.

5. It is ideal to transport sick or recently dead animals to a pathology laboratory for necropsy by trained personnel. Since it is not possible to transport the dead animals in most circumstances, appropriate tissue samples can be obtained by field personnel. Complete tissue and blood samples be obtained from carcasses.

6. Post-mortem examination may be performed in the field or laboratory. Depending on the location and body size of the carcasses one has to decide whether transport to the laboratory facility for examination is possible, or whether the post-

Introduction 9

mortem must be performed in the field. If the field post-mortem examination is unavoidable attention should be paid to the risk of spread of infection to wild or domestic animals through opening of the carcass and available methods for carcass disposal (e.g. pit, cremation). It may be possible and even desirable to transport whole, unopened small animal carcasses directly to the laboratory. Large animals usually have to be examined at the place where they are located.

The necropsy should be carried out within an appropriate interval after the animal's death. In case of wild animals, PME should be carried out soon after the detection or receipt of the carcass. In many situations it is not possible to start a *post-mortem* immediately after death, due to legal constraints or on account of practical consideration i.e. carcass in isolated location or due to health and safety consideration, possibility of zoonotic diseases (in some countries legislation may dictate that a carcass must be checked for notifiable diseases like anthrax before one proceeds to full PME), in such situations, carcass cooling to slow the rate of autolysis should be practiced. The animal's body should be kept at appropriate refrigerated temperature to ensure meaningful necropsy results. Autolysis of the organs occurs with variable speed. Autolytic changes develop quickly in the adrenal medullae, gastro-intestinal mucosa, pancreas, liver, kidney and central nervous system.

When a *post-mortem is performed by self,* collect diagnostic material systematically. The correct selection of material for further examination, and the correct sampling, storage and shipping of material, will increase the quality of results. A written report of the *post-mortem* findings will help the zoo veterinarian to keep track of the disease status of the zoo collection. In areas where rabies infection is enzootic, one should examine all mammals found dead, and particularly those with a clinical history of abnormal behaviour or neurological signs. Consider them as potentially infected until proven otherwise.

In all cases of suspected sudden death, examine the peripheral blood smear to exclude anthrax infection before opening the carcass. Bloody discharges from natural orifices should direct the examiner's attention to the need to exclude anthrax infection before continuing the postmortem examination. The specialist veterinary staff may be legally required to carry out the test.

- Blood samples should be taken by nicking the dependent ear or from the coronary band.

- In wild equids - horses and zebras, wild pigs and carnivores, the smears made from the cut surface of a lymph node (usually sub-mandibular) are examined since anthrax bacilli may not be present within the blood.

- Air dry the tissue and blood smears and then fix them in methanol. Then stain them with polychrome methylene blue, or Giemsa stain for two minutes.

- Examine the smears under oil immersion for evidence of anthrax bacilli.

- If anthrax infection is confirmed, take quick and effective carcass disposal. Regional authorities responsible for disease control should be notified for taking appropriate action.

- If anthrax infection is excluded, the post-mortem examination can be carried out.

Anthrax bacteria are commonly confused with Clostridium bacteria. Anthrax bacteria appear as large (up to 10 μm long) rectangular rods, alone or in chains and surrounded by a capsule. Anthrax bacilli obtained from a recently dead carcass usually do not form spores.

Bacillus anthracis	Clostridium septicum / Postmortem invaders
Sharp, squared ends/Truncated ends	Smaller in size with rounded ends
Pale capsule around bacteria	No capsule

Before starting a necropsy, consider whether the skin or skeleton is important for museum-based studies. If it is, a cosmetic post-mortem is required.

Knowledge of the species of common predators for the animal under post mortem examination in that region is useful. An understanding of the distribution of the wounds that they typically inflict can be a useful aid for identification of cause of death or scavenging.

Note: If poisoning (e.g. plant poisoning) is suspected, the pathologist should be informed before the necropsy is carried out.

Submission and Selection of Animals for Necropsy

It is the responsibility of the clinician to euthanize the animal for post-mortem. Animals submitted should be properly tagged with identification and disposition along with request form signed by hospital clinician or referring veterinarian. The form must include accurate information in regard to age, sex, breed, name of the owner, and pertinent case history. Specially coloured tags can be used (pink for Necropsy, blue for Save, yellow for Disposal Only and green for Cremation). Before performing a necropsy on an animal two important points need to be considered.

Some diseases of wildlife can cause serious illness or death in humans. If the veterinarian or field biologist suspects that the disease outbreak under investigation is likely to be caused by a legally notifiable disease, make immediate contact with the appropriate regional or national veterinary authorities before proceeding with the autopsy.

1. Zoonotic diseases	2. Reportable and Infectious diseases
Salmonellosis, Anthrax (Ungulates and Carnivores)., Rabies (Carnivores, Omnivores and Insectivorous bats), Small tapeworms of Echinococcus spp (Hydatid disease-Carnivores or Rodents-clear cysts in the liver of carnivores), Tuberculosis (Nodules in the lungs, enlarged lymph nodes or thickened intestinal wall- some wildlife population), Brucellosis and Rift valley disease (Wild herbivores, Psittacosis and Newcastle disease (Birds), Tularemia, plague and Hanta virus (Rodents and fleas), Herpes B viral disease (Wild primates) Hendra virus (Fruit bats and insectivorous bats) and Lyssa virus (Australian bat)	Anthrax, Foot and Mouth disease, or Tuberculosis. Spread to other animals through contamination of the environment during the necropsy procedure. Anyone necropsying wild animals should be aware of the typical lesions of these diseases and take extra precautions when decontaminating a necropsy site. Suspicious carcasses should be deeply buried to prevent scavenger access (if anthrax is suspected). If possible, select a site for the postmortem, away from contact with other wild animals and domestic livestock. A pit may have to be dug to dispose of the carcass. So a site where this is feasible should be chosen. If cremation is to be the chosen method of disposal, fire risks should be taken into account.

Specialist examination of the brain is required for diagnosis of rabies infection. Removal of the brain and spinal cord may require experience and specialist equipment, particularly while dealing with very large or small taxa, and those with particularly robust skulls (e.g. Suidae-pigs (family).

Note: In view of the above, person conducting post-mortem examination of wild animals should be aware of risks involved and should wear appropriate protective clothing.

Safety Precautions: Personal safety

Before undertaking any post-mortem examination consider the potential hazards to human health. Some diseases of wildlife can cause serious illness or death in humans. All carcasses should be handled as if they were harboring potentially dangerous diseases (until proven otherwise). Precautions for personal safety should be taken and handled accordingly with due caution. Potential hazards range from toxins on the surface of the animal (e.g. oil) to zoonotic diseases which may be transmitted through cuts, absorbed through mucous membranes or inhaled in the form of dust or aerosol. Zoonoses where knowledge of the natural wildlife host is incomplete include Ebola Virus in Central Africa and Nipah virus in Malaysia. Herpes B virus infection is enzootic among Old World macaques (Macaca spp.), and usually causes minimal or undetectable morbidity in its natural host. Human infections are almost always fatal. A face mask should always be worn when conducting an autopsy on a primate of any species. Conduct necropsy inside a protective cabinet, if possible, This is highly recommended when examining primate carcasses. Any cuts should be washed immediately with a disinfectant soap. Great care must be taken to ensure that all specimens taken from a carcass are collected, stored and transported safely and that there is no risk of the escape of infective material.

Minimal protective clothing includes coveralls, gloves and a mask that covers the nose and mouth and rubber boots. When necropsying a primate of any species, a full face shield, coveralls, and double gloves should be worn. A washable rubber apron is also recommended. Always wear aprons, gumboots and gloves (which should be replaced immediately if damaged). Remove rings and watches. Face masks, which cover the eyes (protective eye wear may be recommended to prevent ocular splash injuries), nose and mouth, are particularly important when examining animals suspected of infections likely to become air-borne during dissection, particularly when aerosol transmission infections are suspected. Wearing a mask is particularly important when performing a necropsy on a primate, bird, or a carnivore suspected of rabies or hydatid disease. Also, all samples should be handled with care and unfixed samples should be placed in leak-proof containers so that dangerous infectious materials do not leak during transport.

Transmission of primate viruses from primates to humans can occurs through bites, scratches, cuts and abrasions contaminated with blood or saliva, contaminated hypodermic needles and contact with infected primate tissues. Herpes B virus infection is enzootic among Old World macaques (Macaca spp.) and usually causes minimal or undetectable morbidity in its natural host. Human infections are almost always fatal. Chimpanzees, and perhaps gorillas, may carry Ebola virus infection and Simian immunodeficiency virus.

Note: Always wear rubber or plastic, disposable, surgical gloves while conducting post-mortem examinations on felids, canids or hyaenids, since they may be infected with Echinococcus spp. tapeworms, the microscopic eggs of which may contaminate the hairs of the tail and hindquarters. These eggs are infective to humans and develop into hydatid cysts, (disinfect them thoroughly after use by boiling, or destroyed by burning). No known disinfectant inactivates the eggs of E. granulosus.

Important Steps to Remember While Conducting Post-Mortem

- Examine the animal in an order. Follow a systematic approach (head-to-tail, system by system--digestive system, respiratory system etc., or any other).

- First consider the history of the animal (where available).

- Note the reported clinical signs, treatment, diagnostic tests, possible differential diagnoses, number of animals involved, etc. (It is better to communicate with the case clinician, where available).

- Consider recent and historical disease problems in the captive animal, free-ranging, in-contact domestic animal and human populations.

- Examine the site where the carcass was found (e.g. evidence of agonal movements, convulsions disturbing the local area; piles of faeces and urine around the hindquarters suggestive of prolonged recumbency).

- Recognize the normal anatomy, normal appearance of organs/tissues and anatomical variation between species. Enlarged adrenals indicate longer-lasting stress prior to death. In primates the adrenals in general are larger than in most other mammals; veterinarians not familiar with this may erroneously believe that normal adrenals are enlarged . The entire gland should be collected with transverse incision for histology.

- Have knowledge of seasonal differences in the body condition and reproductive system which are normal for the species under examination.

- Have knowledge of the expected variations between individuals of the same species kept in captivity and free-ranging. Captive animals are obese unlike free ranging individuals. The free-ranging wild animals have a greater ectoparasite and endoparasite burdens than in those under kept in captivity.

- If euthanasia is done, know the method, route and time of its performance.

- Have knowledge of potential artefactual findings e.g. hypostatic congestion (pooling of blood in organs under the effects of gravity which can be mistaken for pathological congestion, barbiturate crystals from euthanasia solution which can be mistaken for gout, pseudo-prolapse of the anus or vagina as a result of increased pressure within the abdomen caused by gas production after death.)

- Describe lesions/abnormalities accurately.

- Note down both positive and negative findings.

- Keep accurate records, including a unique identifying number for each carcass and for samples from that carcass.

- Keep detailed notes on all findings and procedures for forensic post-mortems, written in non-technical language wherever possible, for use in court. Avoid the use of non-standard abbreviations in permanent records.

- Take photographs for animal identification and illustration of gross pathology, particularly in a Vetero-legal case. Photographs illustrating environmental conditions and the appearance of sick animals and their lesions should be taken.

- Collect tissue samples, swabs for culture and impression smears as tissues are examined and before they become contaminated. Tissue to be fixed should be no more than 3-5mm thick, and thinner if congested, although whole lung may be fixed. Samples for histopathology should not be frozen (since this results in

Introduction

crystal formation and cell rupture).

- If foreign bodies such as bullets or lead pellets are found, collect them , in the presence of a witness (if there is the possibility of legal proceedings), label (including the initials of the label) and store in a safe place.

- Collect tissue samples from all the organs, including those with apparent gross pathology. Direct further investigations at the samples most likely to be important in revealing the cause of death.

- In some circumstances it may be advisable to keep the entire carcass refrigerated in the short term and frozen in the long term, to provide samples in the future if required following the post-mortem examination.

In most cases a full necropsy is essential. Confirm if the animal is dead. (*Some animals can go into a state of hypothermia and hypo-metabolism during hibernation. The ectothermic species (reptiles, amphibians, fishes and invertebrates) can appear dead because of low ambient temperatures and consequently a lowered metabolic rate. Though the traditional indicators of death are useful, it is not always totally reliable. It can be challenged in court. Ectothermic species that are "dead" may still have a beating heart which means they can bleed and even in a mammal or bird sometimes blood comes out of haematoma or well vascularized wound, especially if the animal has clotting defect. Little information is available regarding the determination of death in ectothermic species. These animals exhibit residual cardiac contraction even when clinically dead*). In a carcass if the heart has stopped beating, it is considered as dead.

In the field the carcasses may have to be protected from scavengers including human, pending examination. Wild animals are presented often as decomposed carcasses or parts. How rapidly the carcass decomposes, depends on how it was wrapped, depth of burial, ambient temperature, soil type and the presence or absence of water. Wild animal pathologist may also come across wild animal that has been fixed in formalin, glutaraldehyde or an alcoholic preparations. Occasionally an incomplete examination is necessary either for legal reasons or because of the rarity of the species which means that the carcass is required for the research or display. Necropsy may not be complete for other reasons---because in wildlife work portions of a carcass may be found or only certain derivatives are submitted for examination.

PME can yield data on morphometrics, organ weights and organ/body weight ratios and on gross and histological appearances of tissues. Extensive collection of informatics and pelage may be counter productive in cases where the data are of little or no direct relevance to the main objective of the forensic necropsy. In cases of rare or endangered species, when time is short , request a zoologist with the knowledge of the species that is being examined to attend the necropsy to advise on special morphological features and to record biological (non pathological) data , or may consult a suitable qualified person before doing PME in order to ascertain zoological data, if possible. Experience of and training in such work can be important. It should be done by a specialist veterinary pathologist not even by a veterinary practitioner.

Laboratory post mortem facilities should be housed in a separate room to the clinical facilities. It should be easily cleaned, and have adequate water supply and draining floors. Requirements for air exchange systems, fume cupboards etc. will vary according to the taxa under examination and the pathogen classes which are likely to be encountered.

P.M. Kit

The kit should contain the following items.

1. Protective clothing: Rubber Gloves, Rubber Boots or Plastic Foot Protectors, Rubber Apron (boilable), Coveralls, Mask (to cover mouth and nose) and Eye Goggles or Face Shield.

2. Necropsy documentation: Camera and film, Field notebook.

3. Stains: Stain kits for cytological, bacteriological and fungal examinations (Gram, Ziehl-Nielsen, Diff-Quick®, Hemacolor®, Lacto phenol blue, etc.).

4. Transport materials: An insulated cooler box / Ice coolers, 2 Jars (1 litre)--Leak-proof, break-proof containers, Absorptive packing materials, string and heavy duty plastic sealing tape, sterile buffered glycerin (50%), "Easy blood", 'blue ice' freezer packs (pre-frozen),(culture tubes with sterile swabs; Transport media (without charcoal) with swabs, (for agent isolation especially to facilitate the identification of anaerobic bacteria, or collect samples for bacterial culture, or from samples likely to experience contamination or overgrowth during collection or holding during transport to the laboratory)

Introduction 17

5. Necropsy equipment: Curved knife for skinning, A straight pointed sharp knife for dissecting(including sharpening stone or steel), scissors (small and large),a pair of 15 cm dissecting scissors, a pair of 25 cm rat toothed forceps, 15 cm pointed forceps, an enterotome, String (for tying off loops of bowel), Bone cutter or Axe or hatchet-for opening the carcass and skull, Hack saw or bone saw, a large pair of bone forceps or bone cutting shears, Small and large shears, Chisel and mallet, a spring balance to weigh up to 10kg, Scales (both gram and kilogram),a block and tackle, scalpels and razor blades, Alcohol lamp or gas burner for sterilizing instruments, Plastic ruler or measuring tape, Specimen containers and sampling equipment ,appropriate sterile containers (Petri dishes, plastic bags in which tissues can be frozen), Rigid plastic containers with tight fitting lids (approximately 1 liter), Small vials, tissue cassettes, or tags to identify specific samples, Sterile vials or blood tubes, Plastic bags with closure tops (zip-lock), (Two 8"x10"- for re-bagging fresh tissues, Two- 10"x12"- for re-bagging formalin jars, Twelve 6"x6" for individual fresh tissues and one 6"x9"for fluid tubes), 2 Freezer Packs, (1 small pouch for anaerobic transport media and 1 large pouch for other fresh samples), 2 Jars (1 Litre) with 10% buffered Formalin, 1 Fecal cup (2 ounce), Para film or sealing tape, Aluminium foil ,Sterile syringes (two, 12 ml)and needles (18 / 20 g)—11/2 " for fluid aspiration, Heat source such as a small portable torch and a spatula or flat blade to tear tissues for inoculating microbiological media with minimal contamination, Sterile swabs in transport tubes, Unfrozen ice packs and frozen packs when returning to the lab, Labeling tape or tags, waterproof labeling pens, and pencil, Microscope slides and slide boxes and cover slips for transport, WHO rabies kit (or drinking straw in a small jar of glycerin)

6. Dissection equipment (clearly marked) and kept solely for the purpose. Ophthalmology instruments, hand lens and dissecting microscopes (useful for small mammals where available).Clean and sterilize equipment following use.

7. Equipment: Microscope: A microscope with a mirror for a light source or adapted for car batteries (A field scope will permit assessment for anthrax before opening a carcass), Camera and film, note book. Centrifuge: A portable centrifuge for spinning blood is optimal. (E.g. Mobile spin centrifuge).

8. Fixatives: 10% buffered formalin, 100% acetone for cytology, 70% ethyl alcohol for parasites, Normal saline.

9. Disinfecting materials: Pail and brush, Borax, Sodium hypochlorite (0.5%) (10% Clorox), 70% ethyl alcohol (for disinfecting instruments).5% formalin, 5% Sodium carbonate.

Carcass Handling and Disposal

Diseased wildlife should be handled to minimize exposure of other wild and domestic animals. Carcasses with anthrax or other infectious diseases should be buried (at least 2 m deep, preferably covered with a disinfectant to prevent scavenging). When burying and burning of anthrax infected carcasses is difficult or impossible, due to lack of labour, equipment or fuel, the danger of fire or the sheer number of carcasses, it is possible to limit the contamination of environment with escaping anthrax spores by ensuring that the carcass is protected for 4 days and nights by keeping a guard. The protection from avian and mammalian scavengers is ensured until the changes in pH and the putrefactive organisms within the carcass have destroyed the anthrax bacilli and prevented sporulation. After 4 days in tropics the guard can be withdrawn and left to vultures and hyaenas.

Preparing Sample Containers

All containers, tubes, slides, and bags should be labeled using a waterproof marker. Placing a second label in a plastic bag that is then attached to the container adds further security. For formalin-fixed tissues, a paper label with the animal identification written in pencil can be submerged in formalin with the tissues.

Information to be furnished

- Date
- Geographic location (Park name or nearest town, country)
- Species
- Sex and approximate age
- Tissue Identification (this is not necessary for formalin fixed tissue samples)
- Person taking sample

Introduction

- Animal ID (if available)
- Performing for the Necropsy: General Considerations
- Checking for Anthrax
- Determine if anthrax is present

Different Ways of Approach to Postmortem

Salient features of each method is given.

1. Standard complete necropsy

i. Complete gross dissection of all organ systems including removal of the central nervous system

ii. Histopathology on major organs and major lesions at gross necropsy.

iii. A complete descriptive report of gross lesions and histopathology findings

iv. Morphologic diagnoses and interpretative comment in the report.

v. Summary of results of ancillary procedures viz. Microbiology, virology, Parasitology and toxicology.

vi. Selected gross photographs

vii. Final report in 3 to 4 days.

A preliminary report on gross necropsy findings within 24 hours.
The final report includes: 1) Owner information, animal information, clinician and referring veterinarian; 2) Clinical summary or history on animal written as submitted by the clinician or the owner/farm manager; 3) Clinical diagnosis; 4) Gross necropsy descriptions; 5) Gross diagnoses; 6) Histopathology descriptions; 7)Final anatomical diagnoses and 8) comment if appropriate.

2. Insurance necropsy

i. Similar to standard complete necropsy except the report includes descriptions of normal and abnormal findings.

ii. More complete histopathology.

iii. Photographs of whole body and gross lesions.

iv. A copy of the final report is mailed to the insurance company.

3. Rabies suspect and other biohazardous disease necropsy (Incomplete Necropsy).

i. Similar to standard complete necropsy except the brain and spinal cord segment are removed initially under biosafety conditions for virology and histopathology.

ii. If the rabies virology test is negative, examination of remainder of the carcass that has been held under refrigerated conditions is completed.

4. Diagnostic necropsy
(variation of standard necropsy)

i. This is a modification of the standard necropsy procedure.

ii. The gross dissection will give an overview of organ systems.

iii. A more limited tissue set will be collected and the histopathology will be targeted for definitive diagnosis of the major disease conditions at death.

iv. Similarly Ancillary procedures will be targeted towards major disease diagnosis.

v. The final report will be completed in 1 to 10 working days and will include: 1) Owner information, animal information, clinician and referring veterinarian; 2) Gross diagnoses and/or final anatomical diagnoses; 3) Comment if appropriate.

Cosmetic Post-Mortem on Mammals

In this method the body is examined with less mutilation. Cutting and incisions are sewed together. Body is washed to appear nearly intact. This is done in case of pets and wild animals.

A cosmetic post mortem allows full access to all tissues required by the

Introduction

pathologist. It does not take more time to carry out. It allows the preservation of skins and skeletons for museum-based studies.

Main steps in a mammal post-mortem

1. Weigh the mammal and record its weight.

2. Take measurements (Total length, Tail length, Hind foot length and ear length)

3. Make a small central incision with a scalpel. Make sure to cut between the hair, but not through it.

4. To increase the size of the incision, insert the scalpel blade under the skin facing upwards, so that the skin is cut from underneath. The incision can be increased to extend from the anus to the throat.

    Do not cut through the lips.

    Do not remove any pieces of skin, unless needed for histology. If so, make cuts as small and as cleanly as possible.

5. Peel the skin back from the thorax and abdomen. If possible, rub salt into this skin. Access can now be made to the abdomen.

6. To access the thorax, cut through the cartilaginous ribs close to the sternum and lift the sternum upwards and forwards, so that the ribs can be splayed apart. If access is required to the spinal cord, make an additional incision down the mid-line of the back and peel the skin to the side. As long as cuts are clean and all the skin is there, we can easily sew the pieces together again.

7. For larger mammals, it may be necessary to remove the limbs. This can be done from inside skin, by detaching the scapulae from the muscle holding them to the back and the humerus from the clavicle. For the hind limb, the head of the femur can be detached from the pelvis. Peel back the skin further to allow full access to the detachment points, but do not cut through the skin.

8. If it is absolutely necessary to get access to the brain, make an incision from the crown on the head down to nape of the neck. Peel back the skin and if necessary cut through the bases of the ears. This will allow full access to the cranium, which should be trepanned carefully, so that the skull can be reconstructed after the post-mortem. Alternatively, from the underside of the neck, sever the skull from

the cervical vertebra and peel back the skin from the top of the head and cut through the bases of the ears if necessary. Again access to the cranium is now clear

9. After the post-mortem, if possible rub salt into the paws and face, double bag in polythene and deep freeze. Label the bag clearly with species name, identification name/number and/or date of death.

Autopsy Site

The post-mortem site should be far away from contact with other wild animals and domestic livestock. A cement floor if available greatly facilitates subsequent disinfection. Where feasible, select a site to dig a pit to dispose of the carcass. If cremation is to be the chosen method of disposal, fire risks should be taken into account.

Examining the Carcass and its Surroundings

- Assess the condition of the environment (Note recent weather conditions that could have caused deaths, drought, floods, electrical storm, etc).

- Note the ambient temperature, signs of struggle and condition of the animal.

- Look for any bite wounds or other signs of predation (If wounds are present, look for bruising and bleeding in the tissues near the wounds which would indicate that they occurred before the animal died. Otherwise these wounds most likely were caused from the carcass being scavenged.)

- Look for broken bones, missing hair, broken or missing teeth or other signs of trauma.

- Look for and preserve any external parasites.

- Note the nutritional status of the animal.

- Take weight (if possible) and/or body length and girth. Assess fat stores under the skin and in body cavities.

- Note the amount of fat around the heart and kidneys and muscle mass of the animal.

Introduction

- Note the amount of food in the digestive tract and the condition of the teeth.

- Most carcasses will have some autolysis, but diagnostic tests can still be performed if tissues are properly handled. Handle the autolyzed tissues very gently for histopathology. Hold tissues at the edges only. Cut with a sharp knife or scalpel. Quickly place them in formalin. Freeze or refrigerate samples as soon as possible for infectious disease or toxicology testing. Autolysis can cause many artifacts in tissues that can be confused with a disease process. However, it is always best to take a sample from an area that looks abnormal rather than assume that the change was caused by autolysis. Histopathology will be able to distinguish between true lesions and post-mortem changes.

2 Collection and Preservation of Biological Specimens

In many cases, the pathologist carries the responsibility for the diagnosis of the cause of death except when it is known. The success of laboratory examination depends mainly on the proper collection, preservation and dispatch of suitable materials. Even when the field investigator feels competent to form an opinion regarding the identity of the disease encountered, a complete range of specimens must be taken and submitted, so that a thorough laboratory investigation can be performed. For this, the laboratory should be consulted before performing the post-mortem examination so that specific instructions can be obtained on which specimens should be taken and how they should be preserved and transported. Where possible, a live diseased animal should be submitted to the pathologist at the laboratory. If this is impractical, then a whole, unopened carcass may be submitted provided that it does not get putrefied and useless for diagnostic purposes before it arrives at the laboratory. But in many circumstances, samples of all tissues (that show lesions as well as those that are judged to be macroscopically normal) are collected and preserved and sent to the laboratory. The method of euthanasia may affect specimens submitted to the pathologist. High doses of barbiturates are caustic to tissues and cause crystallization in and on organs. Such changes may be mistaken for early gout, but will also change and mask macroscopic and microscopic lesions. The letter should contain all particulars of the specimens, preservatives used, history of the case and the time of

Collection and Preservation of Biological Specimens 25

animal's death and that of necropsy along with a copy of the post-mortem examination report, the disease suspected and the specific tests required.

Preservation of Specimens

The method of preservation chosen will vary with the type of investigation required and the time that must elapse before the investigation can be made. Before shipping, contact the laboratory. Check regulations for shipping tissue samples. Get proper permits and use the correct containers. Any samples for culture should be kept refrigerated (for parasitology or bacterial cultures) or frozen (for toxicology or virus cultures). Freezing at -70°C is preferable to -20°C (standard freezers). It is best to ship frozen and fixed samples separately. If they must be shipped together, then insulate the fixed tissues from freezing by wrapping in newspapers. Ensure that there is no spillage of formalin, because fixation of frozen samples will make culturing for bacteria or viruses impossible and will alter cells on blood smears or cytology slides.

Refrigeration

Specimens are refrigerated at approximately 4°C (38°F). They will be preserved satisfactorily for a short time. Specimens for transport to the laboratory by post or other means may be packed in leak-proof containers and surrounded by ice in sealed cans or sealed plastic bags. Polystyrene boxes or large thermos flasks may also be used to provide additional insulation.

Dry ice

Solid carbon dioxide is useful but has certain disadvantages. Some pathogens, especially viruses, are partially or completely inactivated by CO_2 vapour. The specimens must therefore be completely sealed away from the refrigerant. Delay in shipment resulting in complete evaporation of the CO_2 block may be followed by a drastic rise in temperature which could be much more detrimental to the survival of many pathogenic organisms. If dry ice is used as a refrigerant the outer container, must not be airtight, as there is danger of an explosion if dry ice sublimates in an airtight box.

Deep-freeze

Freezing in a deep-freeze cabinet at a temperature of $-20°C$ is satisfactory for some but not all pathological specimens. A deep-freeze will not be appropriate for specimens destined for histology nor for such organisms as Vibrio foetus, Leptospira spp., or Toxoplasma spp., all of which are damaged by freezing. Serum, however, is well preserved. Whole blood should not be frozen. Frozen samples must be shipped in insulated containers and by express carrier. Pack specimens in dry ice or with ice blocks. Seal container to prevent leakage. Include proper permits and animal identification.

Diagnostic Materials

- **Blood smears**

Thick blood smears are prepared for examination of microfilaria (peripheral blood smear-venous blood) or trypanosomes. For all other examination blood smear should be thin (One should be able to read newsprint through them). In all cases, the smears should be dried quickly, labeled with a diamond pencil or grease pencil and wrapped or boxed to prevent damage by insects. Blood smears are fixed immediately in methanol. Thick blood smears should not be fixed. It is of some advantage to place the thick and thin smear on the same slide. Blood smears are best made from venous blood collected from a live animal but useful smears can be made from blood collected from an animal which has recently died. Peripheral capillary blood maybe desirable for making smears for the detection of microfilariae, which are often cyclical in their appearance in the peripheral blood.

Blood films for microscopic examination should be thin. It is best to collect the blood from the tip of the ear in the case of a living animal or of an animal suspected to have died of anthrax.

Procedure

1. Clip the hair from the tip of the ear, if necessary.

2. Wipe away hair and dirt from the clipped area with a little dry cotton wool. Swab with methylated spirit and let it dry.

Collection and Preservation of Biological Specimens

3. In a living animal, make a puncture at the tip of the ear with a pin that has been sterilized by flaming, while in a dead animal snip out a little piece of the skin at the tip and allow the blood to ooze from the wound.

4. Bring the surface of a slide in contact with the blood droplet or transfer the droplet to the slide by means of the edge of another slide. In either case, the blood should be deposited at a point on the slide about 1 cm from one end, while the quantity of blood removed from the wound should be just sufficient to spread within the middle half of the slide, leaving the ends blank.

5. Take another clean slide with a straight smooth edge at one of its ends for use as a 'spreader'. Place this edge just in front of the drop, holding the slide firmly, at an angle of about 45° on the lower slide, which may be placed on a table or some other flat surface. Now bring the lower end of the spreader slide in contact with the drop of blood, which will spread along the end of the slide by capillary attraction. Glide forward the upper slide evenly at a uniform pace so as to spread the droplet in a thin uniform film on the surface of the lower slide . From animals suspected to have died of anthrax prepare thick smears (better to make two thin and two thick smears in each case) to send to the laboratory. While making films, protect the slides from the direct sun rays, especially in summer months, to avoid rapid drying of the blood and the consequent formation of artifacts.

6. Dry the films in shade by waving them in the air.

7. In the case of films to be examined for protozoa or blood changes, immerse the slides in acetone-free methyl alcohol for about 10 minutes and then allow them to dry. Smears to be examined for bacteria should be fixed by repeatedly passing the slide, film upwards, through a flame until the glass is just uncomfortably hot when felt with the back of the hand.

8. Mark the slides with a grease pencil or fix labels on them to indicate the nature of the smears.

9. Place the slides back to back and wrap in a clean paper.

10. Enclose a brief description of the smears while packing for dispatch.

- **Pus**

Prepare a thin smear by spreading the pus evenly on a side, with a sterile scalpel or prepare the smear as indicated below.

Place a small quantity of the pus on the middle of a slide. Place a second slide on the first one so that the pus is held between the two slides. Press gently the two slides together so as to spread the pus. Hold the opposite ends of the two slides in each hand and draw them apart, taking care not to lift one from the other (as this will cause the formation of lumps and bubbles). If a thin smear is not obtained in the first attempt, the slides may be placed together again and the process repeated. Dry the films by waving in the air, fix over a flame, label, pack and dispatch.

- **Secretions and excretions**

These can be collected in clean vessels and sent under refrigeration to the laboratory for detailed examination. It is always better to have a Styrofoam (Thermocol) box filled with ice to carry the various biological materials for culture and isolation of organism and for sero-diagnosis. The discharge can also be preserved using 10% neutral formalin (2 or 3 ml for 3-5 ml of discharge).Smears can also be prepared immediately on clean slides and fixed rapidly by heat. These should be wrapped properly and kept. For oozing lesions and cavities containing pus or other materials (as seen in TB of lungs) - a small drop may be collected on a clean slide and uniformly spread using a tooth pick or swab and fixed immediately in heat. In suspected cases of TB it is better to have material scraped out from the lesions (after evacuating the contaminated pus) collected on a clean slide and fixed.

- **Impression smears**

Impression smears of fresh cut organs or altered surfaces are not common practice at *postmortems*.

Make a clean cut with a scalpel blade across the surface of the abnormal area of the tissue. Grasp the sample firmly with forceps, and place the cut surface down. Blot the cut surface of the sample across a paper towel or other absorbent surface until no blood or fluids are evident. Then gently touch the blotted surface in several places on clean slides. Air dry slides. If other fixation is necessary (e.g. heat fixation for acid-fast stains), it can be done after air-drying or cut out a small piece of the organ, hold it with forceps and rub its cut surface on the middle third of the slide. If fluid accumulates on the cut surface, transfer a very small quantity of the fluid to a slide with a sterile scalpel and then spread it. If the cut surface is very dry, scrape it with a sterile scalpel and spread the scraping evenly on a slide. This may be done with

Collection and Preservation of Biological Specimens

the help of normal saline, if necessary. Caseating nodular and calcified lesions may be treated in the manner described for pus. Wrap the slides in paper or a box to exclude insects and dispatch them to the laboratory. Two sets of impression smears are made from liver, spleen, lung and rectum in addition to the organs with pathological changes. This technique is very useful to detect the presence of bacteria, yeasts or protozoa. Tissue phases of parasites such as *Atoxoplasma* spp, *Toxoplasma* spp., *Plasmodium* spp, *Haemoproteus* spp., *Leucocytozoon* spp, and *Trypanosoma* spp. are mostly readily identified in impression smears of liver, spleen and lungs. Immunofluorescent staining for *Chlamydia* spp. can be carried out in specific laboratories on the impression smears of these organs in all *post mortem* examinations of suspected cases in reptiles and birds especially within the families of *Columbiformes* and *Psittaciformes*. It is easier to diagnose the cell-type of lymphoreticular and haematopoietic neoplasms from impressions of liver, spleen and bone marrow than from histology.

- **Blood samples (unclotted)**

Blood should not be frozen.
The following preservatives are used for preservation of blood sample.

1. Sodium fluoride 20 mg/ml of blood

2. Solution containing 10g of sodium citrate and 200 mg of mercuric chloride dissolved in 100 ml of distilled water. One drop of this solution is sufficient for each ml of blood. The appropriate amount of fluid is taken into the clean dry glass bottle; the fluid is dried so that the salt remains sticking in the bottle. Then bottles are cooled at room temperature and the sample of blood to be preserved is taken into these bottles.

3. Small bottles containing dried ethylenediamine tetra-acetic acid (EDTA) or potassium oxalate are used to collect unclotted blood samples from animals which have just been killed. The bottle should be thoroughly but gently agitated as soon as possible after collection to distribute the anti-coagulant evenly. Vigorous shaking can cause haemolysis. Unclotted blood samples should be stored on ice or in a refrigerator. Blood treated in this manner may be used for making blood smears or for packed cell volume (PCV) and haemoglobin estimation. If blood is collected from the cut throat of an animal, care should be taken to avoid contamination with regurgitated rumen contents that may emerge from the cut oesophagus. It is better to collect the blood sample with a needle. If this

is not possible, the skin should be cut to expose the jugular vein and carotid artery. These structures should be cut but the oesophagus on the left side of the neck should be avoided. For mineral analysis, special bottles or tubes are used to collect blood samples. These contain an anticoagulant and after mixing with the blood, are stored in a refrigerator.

- **Diagnostic Materials for Serology**

Serum should be placed in sterile tubes then stored and shipped frozen.

Serum samples for serology

Blood for serum production should be collected as aseptically as possible in sterile glass or plastic vials. In emergencies, blood for serum separation can be collected in plastic bags or even plastic gloves. Serum should be placed in sterile tubes then stored and shipped frozen. After collection, the blood is allowed to stand for a few hours at room temperature so that it will clot. The serum is then pipetted into a sterile tube and refrigerated or deep-frozen. If neither of these methods of preservation is available, serum can be preserved by adding 0.5% phenol or 0.01% merthiolate. Normally, blood for serum separation is collected from the veins of a live or recently killed animal. If the collection is made with a syringe, care should be taken not to apply too high a vacuum when withdrawing the sample because this will cause haemolysis, which may render the sample unsuitable for certain tests. The red cells of some species are very fragile and in these cases it is often difficult to obtain a clear serum sample. Haemolysis can also occur due to excessive agitation of the blood sample, freezing, contamination with water, over-heating (as in the glove box of a car), and bacterial contamination. In the field, blood may have to be collected from an animal which has been dead for some time and in which the blood has already clotted. In recently dead animals, the right heart usually contains plasma clots (yellow/tan material), un-clotted blood, or clotted blood. The heart is a good source of a mixture of blood clot and serum. Plasma or blood should be removed into a test tube and left undisturbed for approximately 30 min to encourage clot formation, then centrifuged at approximately 2000 X G for 20 min. When a centrifuge is not available, serum can still be obtained by letting the clot or blood cells to settle. Separate the serum or plasma from the blood cells, divide into two aliquots, transfer to vials, and then refrigerate or freeze (-20°or -70°) until transported to a laboratory. Serum vials should be labeled with the species, animal ID, date, and owner (e.g., country and park) using a waterproof marker. Such mixtures can be separated by centrifugation or sedimentation in a refrigerator. Prompt refrigeration is desirable for all blood samples taken from dead animals since bacterial contamination is certain. Some immunological tests do not require blood serum (cervico-vaginal mucus agglutination tests for vibriosis, brucellosis and trichomoniasis). Mucus is collected by pipette or tampon, placed in a sterile tube and refrigerated. As an alternative to collecting serum for serology, a minimum of 50g of lung tissue can be collected and frozen. If a centrifuge is not available and blood is obtained from a live animal or a dead animal whose blood has not yet clotted, remove whole blood into a blood tube, let the blood clot with the tube inverted (rubber stopper down), then turn the tube up and very carefully remove the stopper with the blot clot attached, leaving the serum in the tube. Serum or heparinised haematological samples should be collected prior to euthanasia. This may be helpful in diagnosis of endocrine disorders or viral infections. Routine haematological tests may also be performed on these samples.

- **Histology**
Tissue Sampling

The material collected for histological examination should be properly fixed when it is in a fresh condition before being dispatched. The aim of fixation is fix the tissues in, as normal a condition as possible and to prevent post-mortem changes in them. The quantity of the fixing solution used should not be less than 10 times the volume of the material. Cut the tissue into thin sections, each around 0.5 cm in thickness so that the fixative may penetrate and kill the cells quickly. The most commonly employed fixing reagent for general histological work is a 10% solution of formalin in normal saline, which not only fixes the tissues in 48 hours but also preserves them. Handle samples carefully by grasping at the edges without scraping surfaces of tissues or compressing them with forceps. Most tissues do not need individual labeling. If a tissue needs special labeling (e.g. a specific lymph node), place it in a different container (or tissue cassette) or attach a piece of paper to the tissue with string or a pin and label the paper or container with pencil or waterproof marking pen. Formalin-fixed samples can be kept at a cool room temperature until shipped. Formalin-fixed tissues should be fixed for at least a week. If the formalin solution is not freshly prepared, formic acid will be formed. A layer of pieces of marble at the bottom of the container or bottle will bind the formic acid to a precipitate, keeping the formalin neutral. This prevents the formation of "formalin pigment" in histological specimens in improperly fixed tissue samples.

Common Fixatives

1.10% formalin (4% formaldehyde)
40% Formaldehyde   100.0 ml
Distilled water   900.0 ml

2.Formal saline
40% Formaldehyde 100.0 ml
Sodium chloride   9.0g
Distilled /Tap water 900.0 ml

3.Neutral Buffered formalin
40% Formaldehyde 100.0 ml
Distilled water   900.0 ml

Sodium Dihydrogen phosphate monohydrate.......4.0 g
Disodium hydrogen phosphate anhydrous..........6.50 g

Keep the tissues in a fixative for 24-48 hours at room temperature. The volume of the fixative added is 10 times the volume of the tissues.

Collection of Fresh Specimens

Brain: Collect appropriate specimens. Place 1"x1"portions of cerebellum, brainstem and cerebrum into each of 2 different labeled sterile specimen bottles / zipper lock bags for microbiology and toxicology. The remainder of the brain may be placed in the largest formalin jar. For best fixation slice the brain at every 0.5 cm most of the way through leaving about 0.5to1.0 cm of tissue unsliced along the ventral portion to keep the parts together. Collect $1/4 - 1/_2$" piece of ear skin(bovine) into plastic vial containing phosphate buffered saline, 2" x 3" Adipose (fat) piece, 1" x 2" Liver piece, 1" x 2" Kidney, and Intact eyeball in labeled bottles, Collect approximately 2 teaspoons of colon contents for parasitology in labeled fecal cup (do not fill more than 1/3 full) and approximately 2 tablespoons of simple stomach or rumen contents in labeled bottle. If possible, freeze stomach contents.

Additional fresh tissues

Collect additional fresh tissues for microbiological testing, depending on disease presentation. Ex. collect lung if respiratory disease is suspected. Collect multiple samples if lesions are noticed in different parts of the same organ system, such as small intestine, caecum and large intestine. Collect small lymph nodes as a whole. Collect $^1/_2$ to 1" pieces from larger lymph nodes (retropharyngeal, bronchial, mesenteric or peripheral lymph nodes) for microbiology. Place all fresh tissues in labeled bottles.

Collection of Fixed Specimens

Nervous System: 1. Brain: Put the remainder of the brain, into the formalin jar(s) if not already done. It may be necessary to divide it for fixation and shipping into two jars. 2. Peripheral nerves: Collect a segment (1")of a peripheral nerve, such as the sciatic Nerve. 3. Spinal Cord: Collect one or more sections, from cervical, thoracic, and lumbar areas, by disarticulating or sawing through the spinal column at various levels.

Other tissues: Cut thin ($^1/_4$"/0.5 -1.0 cm) sections of each of the following tissues and place in the other jar of 10% formalin. Include sections that show both normal and abnormal appearing tissue where lesions are recognized.

Gastrointestinal (GI)System: 1.Tongue, 2.Liver, 3.Esophagus, 4.Gall bladder, 5. Stomach (sample each compartment, if ruminant), 6.Pancreas, 7.Duodenum, 8. Jejunum, 9. Ileum, 10. Caecum, 11. Colon.

Lymphoid System: 1. Bone marrow (Break a long bone or rib, exposing marrow, and include piece with exposed marrow. Adults: collect near end of long bones; large livestock: ribs work well.), 2. Mesenteric, trachea-bronchial and peripheral lymph nodes, 3. Thymus, 4. Spleen.

Respiratory System: 1. Larynx, if indicated, 2. Trachea, 3. Lung: Include at least sections of cranio-ventral, middle and dorsal portions, but include representations of lesions and normal sections, if gross lesions are recognized.

Endocrine System: 1. Adrenal gland, 2. Parathyroids, 3. Thyroid, 4. Pituitary.

Urinary System: 1. Kidney, 2. Urinary Bladder,3. Ureter, if indicated,4. Urethra.

Reproductive System: 1. Representative samples of male or female genital tract and gonads. Include fetal and fetal membrane samples, if indicated, for pregnant female.

Musculoskeletal System: 1. Skeletal muscle, 2. Bone: if indicated, 3. Diaphragm.

Cardiovascular System: 1. Pericardium, if indicated, 2. Myocardium, 3. Aorta

Integumentary System: 1. Skin, 2. Nail, hoof, horn, if indicated.
Special Senses: 1. Eye, 2. Ear, if indicated.
Preserve samples (tissues, parasites etc.) for further testing and future reference. Always collect a full spectrum of samples at the initial examination and store appropriately. If further samples are needed subsequently, they can be sent from the stored ones. Each formalin container should be tightly closed. Place the post mortem report and any supplemental history or pictures. Send them to the laboratory.

In instances where the carcass is extremely small, such as embryos, nestlings or very small adult animals, the entire carcass may be submitted for histological examination. This is best accomplished by opening the body cavity, gently separating the viscera and fixing the entire carcass in formalin solution. When transporting a carcass or pathological sample to a laboratory for analysis, attention should be paid to temperature control in transit. Frozen tissue specimens or carcasses must be packed with sufficient ice to keep them frozen until arrival at the laboratory. Refrigerated or frozen specimens should be packed in a sturdy, insulated box, preferably in a leak-proof seal bag, and shipped to the laboratory by a private courier service, which guarantees same or next-day delivery to the laboratory. Ensure that sufficient refrigerant be packed with the specimen and that it be adequately insulated to ensure that it will remain cold (or frozen) until it is received by the laboratory personnel. Take the following precautions.

- Use insulated containers and ice packing of frozen samples.

- Do not send the samples when postal delays may be expected (e.g. weekend, public holidays, strikes).

- Consult postal authorities regarding local regulations governing the postage of pathological samples (labeling, courier, container type etc.).

Microbiology (Bacteriology and Virology)

Any specimen for bacteriological examination must be taken as aseptically as possible. The instruments used should be boiled for 15 minutes before use. If facilities for boiling do not exist, contamination of instruments can be reduced by swabbing these with alcohol and then flaming until red hot. They should be cooled before use. Alternatively, if the investigating laboratory is nearby, a large specimen of tissue (>5 cm^3) may be collected, placed in a clean, plastic container, refrigerated and transported rapidly to the pathologist. The material must be collected with aseptic precautions and dispatched in sterile containers to prevent contamination from extraneous sources. In the case of dead animals, the specimens should be taken soon after death, for otherwise putrefactive bacteria invade the tissues and render them unsuitable for examination.

Solid tissues such as liver, spleen and kidney, may be forwarded, if fresh, without a preservative, on ice, when the examination is to be carried out within a short time of their collection. If however, the examination is to be delayed by a few days, it is preferable, especially in summer months, to preserve the tissues in a 25% glycerin saline. Liquid material such as inflammatory exudates, heart blood and cerebrospinal fluid may be taken either in sterile swabs or sealed in pipettes. If peritoneal fluid is required, an area on the abdominal wall is seared thoroughly with a hot spatula and a sterile forceps is used for holding the cavity open and sterile pipettes for drawing the fluid. Similarly, for collecting heart blood, the surface of the organ should be well seared and a sterile pipette inserted into it for drawing the blood. For taking swabs of the contents of a closed abscess in an animal, clip the hair from the area and paint it with tincture of iodine. Open the abscess with a sterile scalpel and take swabs from the wall as well as from the contents of the abscess and keep the swabs in sterile tubes. Refrigeration is the best means of preserving tissues for bacteriology. Specimens of heart blood, pericardial, cerebrospinal, and joint fluids should be placed in sterile bottles and stored and transported under refrigeration.

Specimens of hair and skin scrapings for mycological examination may be placed in a paper envelope or glass vial. Some animal mycoses are transmissible to humans. Samples also can be taken with a sterile swab, sterile syringe, or by placing a large (3 cm^2) section of tissue in a sterile container (the center of the tissue will be uncontaminated).

Take samples that contain abnormal areas . Small animal carcasses may be submitted intact under refrigeration. Sterile cotton or calcium alginate swabs may be used to collect samples of faeces, pus, heart blood or other body fluids. Care should be taken to see that the saturated swab does not dry out because slow drying is lethal to many organisms. Swabs may be transported under refrigeration or immersed in semi-solid bacteriological transport medium. When taking samples from infected tissues, select an area near the edge of the affected tissue where live organisms are most likely to be found. If no abnormal areas are present, take standard tissue samples of lung, liver, kidney, spleen, tonsil and intestines. Keep samples moist with sterile transport media, sealed in a sterile container and cold. If refrigeration is not available, samples can be placed in buffered glycerin. Smears of pus or infected tissues also are useful and can be air-dried and shipped with cultures. Use Transport media to facilitate the identification and isolation, especially of anaerobic bacteria, or collect samples for bacterial culture, or from samples likely to experience contamination or overgrowth during collection or holding during transport to the laboratory.

Whenever, anaerobic cultures are indicated it is appropriate to inoculate special transport media to avoid bacterial contamination and overgrowth when samples can not be shipped promptly, The appropriate tissues for anaerobic culture are intestine or muscle when Clostridial infection is suspected. This media will also support aerobic and fungal culture. Amies transport media may be included for taking swabs of tissues for which aerobic or fungal culture, without anaerobic culture, is most appropriate (for example, lungs). When sampling from solid tissues, sear the surface of the tissue sample with a heated blade or flame with alcohol and then make a stab incision with a sterile scalpel blade before inserting a sterile swab. When sampling hollow organs such as loops of bowel, it may be necessary to open a segment with a clean scalpel or scissors and swab the interior. If the carcass is severely autolyzed, culture may not be worthwhile. In such situations, consult laboratory bacteriologists before. To collect wound discharge, the wound should be thoroughly cleaned with warm water and soap, and sterile non-antiseptic cotton wool or gauze dressing applied. The material should be collected after about 24 hours by inserting a sterile swab underneath the dressing. This is usually done in living animals.

Virus material for examination may be forwarded in 50% glycerol saline, or preferably in a medium containing equal parts of pure glycerol and M/ 25 buffered phosphate. As the viruses are generally short-lived, it is particularly important that the material should reach the laboratory in the shortest possible time after collection,

Collection and Preservation of Biological Specimens

preferably over ice in a thermos flask. Scabs from pock diseases may be forwarded in a dry specimen tube without the addition of preservative. In the absence of refrigeration facilities, specimens for virology can be stored and submitted in 5 to 10 volumes of sterile 50% buffered glycerin solution.

Sample Collection and Submission for Rabies

In areas where rabies infection is enzootic, all mammals found dead, and particularly those with a clinical history of abnormal behaviour or neurological signs, should be carefully examined and considered as potentially infected (rabies) until proven otherwise. In the case of carnivores suspected of rabies, the head should be severed and preserved.

To remove the entire brain
Separate the skull from the neck at the junction with the vertebra. Remove the skin from the top of the head then the top of the skull. Remove the brain and cut it in to two halves. Preserve one half in formalin and split the other half into containers for virology and toxicology.

Removing the brain if rabies is suspected
If rabies is suspected, extra precaution should be taken while removing the brain. The person removing the brain should wear a face mask and eye goggles. If the animal concerned is relatively small, the entire head can be shipped under refrigeration in a sealed container labeled 'Rabies suspected'. Freezing is not recommended. If refrigeration is not practical, the brain can be removed from the skull and cut in to two halves lengthwise in the midline. One half should be sent in 10% buffered formalin for histology and the other half in sterile 50% buffered glycerin solution to test for rabies.

Unstained smears from the hippocampus, fixed in methyl alcohol, can also be sent. Large animals can be treated similarly, but if it is impractical to send the entire head under refrigeration, the brain may be removed and either refrigerated or preserved as above.

The safest procedure is to remove the head and insert a drinking straw through the hole at the base of the skull where it is attached to the neck. The straw should be inserted in the direction of the eye. Pinch the base of the straw and remove the straw with the brain sample. Then cut the straw (with the brain sample still inside) into 1 cm lengths and drop the sample either into sterile 50% buffered glycerin or 10% buffered formalin. Although this procedure is very safe, it only allows testing for rabies. So if the samples are negative for rabies, no other brain tissues are available to test for other diseases.

Small Animals and Wildlife
Send entire head from companion animal species and small wildlife.

Livestock
In some positive cases, the rabies virus may be detected unilaterally only. By submission of half the brain, on a longitudinal section, the diagnosis of rabies maybe missed. Do not submit entire heads from livestock species. If the testing is required to rule out rabies and BSE/Scrapie, the brain stem and cerebellum can be removed through the foramen magnum, using the following guidelines:

- Sever the head between the occipital bone and the atlas (first vertebra).
- Insert a sharp knife with a long, thin blade into the foramen magnum, being certain it is within the Dura mater.
- Using a circular cutting motion (similar to coring an apple) carve out a plug of tissue by making deep circumferential cuts around the inside of the bony cavity and dorsally until the knife tip hits the dorsal aspect of the cranium.
- Remove the knife, and using a long pair of forceps, reach into the foramen and pull out the excised chunks of CNS. A long-handled spoon, such as an ice tea spoon, will facilitate this step. The first tissue to exit will be the brainstem sample, which should be relatively intact. Submit the brain stem. With continued effort, pieces of the easily recognizable, highly convoluted cerebellum will be removed. Remove as much cerebellum as possible. When cerebellum is submitted in many pieces, rabies testing personnel will make a judgment that they have at least 1/3 or more of the entire cerebellum in order to call the testing conclusive.
- Do not chemically fix the tissue. Preserve by refrigeration only. DO NOT FREEZE.
- The cerebellum and brainstem samples must be placed in a small, crush-resistant plastic canister or tub, then sent to the Rabies lab.
- Tests are also conducted for the following diseases.

Collection and Preservation of Biological Specimens

Cattle - BSE ,Horses and Camelids – Arbovirus ,Sheep and Goats – Scrapie and Deer – Chronic Wasting Disease.

Storage or Submission of Samples

Preparation of 50% glycerol in phosphate-buffered saline (PBS) pH 7.5 1.

1. 1. Put the following chemicals in a flask

NaCl	8.00 g
KC1	0.20 g
KH2PO4	0.12g
Na2HPO4 (anhydrous)	0.91 g

2. Add distilled water to 1000.0 ml
3. The PBS solution may be sterilized by autoclaving at 8-kg pressure for 35 minutes.
4. Add equal quantities of PBS and glycerol. The glycerol should be of analytical reagent grade and neutral. Prior to its being added to PBS, it should be autoclaved at 5-kg pressure for 10 minutes.

Sterile Buffered Glycerin (50%)

For transporting tissues for culture when refrigeration is not available.

To make sterile buffered glycerin, mix glycerin with an equal amount of buffer composed of:

a. 21 g citric acid mixed in 1000.0 ml distilled water

b. 28.4 g anhydrous sodium phosphate in 1000.0 ml distilled water

Mix 9.15 ml of A and 90.85 ml of B

Mix 100 ml of buffer with 100 ml of glycerin. Then sterilize in small tubes to take into the field.

"Easy Blood"

For transporting DNA from blood cells for genetic studies when refrigeration is not available.

Also can be used to preserve DNA for longer periods of time if refrigerated or frozen.

1.2 g Tris HCl

3.7 g Na2 EDTA

2 g sodium dodecyl sulfate (SDS)

Add water to 100 ml

Specimens for Parasitology

Make at least 3 blood smears on clean glass slides.
Fix approximately 2.0 g feces in 70% ethyl alcohol or formalin.
Parasites are extremely common in wild animals and rarely cause clinical disease or death.

Dung samples

Collect the faeces from the rectum of a dead animal. Make an attempt to estimate how long the animal has been dead. The best specimens are collected from the ground as soon as they have been passed. Tightly pack the collected faeces in an universal bottle to the brim to exclude all air, to avoid hatching of nematode eggs. Samples can be refrigerated to further delay maturation of the nematode eggs if egg counts cannot be carried out promptly. If the dung samples are to be preserved for more than a few days, a piece of cotton wool, soaked in paradichlorobenzene,may be placed at the bottom of the Universal bottle and covered with a piece of gauze. The faeces are then placed loosely on top. If the faeces are semi-solid, holes must be pierced in the mass to allow the vapour of the paradichlorobenzene to penetrate. Specimens treated thus can be stored for several weeks but since the vapour kills the eggs they cannot be used for larval studies. If disease due to internal parasites is suspected in a herd, dung samples should be submitted from at least five animals showing signs of disease and from at least five normal animals. The young and adolescent sectors of the population are the most likely to be affected.

External parasites

Many external parasites tend to leave a dead animal as it begins to cool. Hence an attempt should be made to collect some representative specimens either before or as soon after death as possible. Ticks, fleas, lice and mites can all be collected into small tubes containing 70% ethyl alcohol. Lice and their eggs may be attached to the hairs. Mites may require a skin scraping. A label, written in pencil, should be placed inside the tube and should note the part of the body from which the parasites were collected. Ectoparasites can be collected from small mammals and birds by placing the entire carcass in a plastic bag containing a piece of cotton wool soaked in ether or chloroform. Stupefied parasites can easily be shaken or brushed from the body and transferred to a tube of 70% ethyl alcohol with a small paintbrush. Ticks may be required alive if they are to be examined for infectious agents. Not all ticks live on the host. A search should be made of nest burrows, birds nests, roosts, bat caves

Collection and Preservation of Biological Specimens

etc. where the hosts may spend the day or night. Soft ticks (Argasidae) are often found in cracks in rocks and under tree bark in such sites. These may carry infectious agents, some of which are pathogenic for humans, so caution must be exercised to avoid being bitten. Live ticks should be placed in a small glass or plastic vial, the mouth of which is covered with a piece of tight weave cloth or plugged with cotton wool. A strip of blotting paper moistened with few drops of water should be placed inside the tube. For transport, the vials should be placed in a cardboard box, loosely packed to allow ventilation. Engorged ticks travel best. The number of ticks in each tube should be limited and specimens from different animals or sites should not be placed in the same tube. While a record of the presence or absence of ectoparasites is of some value, an account of the numbers present is much more useful.

Internal parasites

If an investigation of a parasitic disease is undertaken contact the parasitologist before sending material, so as to receive special instructions concerning preservation and numbers of specimens required. Collect the internal parasites as soon as possible after the host dies or is killed since these break up rapidly after the death of the host . At least 50 specimens of each parasite should be collected. In the case of tapeworms, collect the whole worm undamaged. If it is impossible to remove the head of a tapeworm or thorny- headed worm from the host tissues without damage, leave the worm attached and cut out the piece of tissue. Worms occupying nodules should be teased out and fixed free. The larval tapeworm cysts should be fixed as a whole. The worms should be washed in normal (0.9%) saline solution to remove mucus and host debris, then fixed in a large volume of fixative. Do not place the worms in the bottle and then pour the fixative over them.

Nematodes (roundworms)

Specimens are dropped into hot (steaming) 70% ethyl alcohol, which kills them instantly. The dead worms should be transferred into a storage bottle containing cold 70% ethyl alcohol and 2% glycerin to prevent drying out should the alcohol evaporate.

Trematodes (flukes)

Trematodes should be washed in saline and then transferred to cold 10% buffered formalin and shaken vigorously. This causes muscular fatigue in the worms and prevents them from contracting excessively. It is sometimes useful to flatten one or two flukes (not conical amphistomes from the rumen) between glass slides. The slides are held together by elastic bands and dropped into the fixative solution.

Cestodes (tapeworms)

Cestodes should be fixed in 10% buffered formalin by suspending the worm by its posterior end and dipping several times into a jar of fixative. The weight of the worm will keep it extended during fixation. A fixed tapeworm should be fully extended with all segments visible and should include the head. Acanthocephalans (thorny-headed worms)are fixed in 10% buffered formalin ensuring that the proboscis is everted. The specimen should be compressed between two glass slides until it everts its proboscis, after which the worm should be fixed.

Labeling

Tubes should be filled to the brim with fixative, any air bubbles tapped out and the cap screwed on tightly. Each bottle should contain the following information.

Host: Full common name/Latin name

Position of parasite in host: Precise distribution in host

Host locality: Name of the area and map/GPS reference

Date of collection

Fixative used

Collector's name

Sample number

Additional information : Include a description of the clinical signs of disease, morbidity, mortality and relevant ecological data.

Address separate letter to the laboratory giving these information.

Specimens for Toxicology

1. A successful toxicological examination requires appropriate specimens and a thorough history, including clinical signs, treatment, necropsy findings and circumstances involved. If a known poison is suspected, a specific analysis should always be requested.

2. In all vetero-legal cases, maintain an accurate record of all the persons keeping the custody of the material from the time of collection of sample till the final analysis in the laboratory.

3. If feed or water is suspected as the source of poisoning, samples of these and any descriptive feed tag should accompany the tissue specimen. A representative feed sample should be submitted from the lot involved in the poisoning.

Collection and Preservation of Biological Specimens

4. Pack the specimens individually in the glass or plastic containers to prevent contamination by lead in soldered joints of cans (zip lock bags are preferable). Metal tops on jars should also be separated from the tissue by a layer of plastic or other impervious materials. Label the containers with all the information necessary to identify the specimen and if mailed, confirm to postal regulations.

5. No preservative should be added except in the case of nitrate poisoning. If a preservative is necessary because of distance from the laboratory, packing in dry ice or ethyl alcohol (1 ml/g of tissue) is advisable. But in the latter case, a specimen of the alcohol should also be sent. Ingesta and tissue should be kept separate, as diffusion is likely to occur between the two.

6. Preserve the materials in 50% of ethanol (1 ml/g of tissue). Tissues and fluids for analysis should be as fresh as possible, kept in refrigerator or preserved chemically. Packing with ice is preferred. Adequate refrigeration is of special importance when submitting body fluids and materials for nitrate analysis, as these salts are rapidly metabolized by microorganisms and only low or insignificant levels may be found on analysis. Refrigeration prevents microbial growth and helps to ensure that the salts are preserved.

7. In some case, if an adequate amount of involved feed is available, some of it may be fed to experimental animals in an effort to produce the signs and lesion observed in the field cases.

8. Samples for toxicological examination should not be washed during collection as washing may lead to the dilution of the incriminating toxic material. Submit generous blocks of liver and kidney, and specimens of stomach and intestinal contents. Blood and urine may also be useful. Specimens should be placed in clean leak-proof jars without any chemical preservative. Aluminium or plastic may interfere with tests for some toxins. Refrigeration or freezing in transit is essential.

Materials for Detection of Poisons

Materials to be collected in suspected cases of poisoning are as follows:

a. 1000-1500 g of stomach /Rumen contents and stomach walls
b. 1000-1500 g of intestinal contents / faeces

c. 1000-1500 g of liver without gall bladder
d. 1000 g of spleen
e. Urine-1 litre in a separate bottle and put thymol as a preservative
f. Kidney-one
g. Brain
h. Fat
i. Aqueous humor or intact eyeball
j. Skin
k. Heart blood (collected into heparinized tube/ blood collection tube)

In survival cases, the following materials may be sent for analysis: Stomach wash, stomach contents, vomitus, blood, urine, faeces, water and feed. Where the poison is suspected to be consumed by inhalation, parts of small intestine with its contents, liver, one kidney, lung, heart and brain tissues are sent. Uterus and foetus are useful in suspected cases of abortion. Burnt bones ashes should be preserved for analysis, if dead body has been cremated. The skeleton or the remnant bones are important materials for analysis in cases of exhumed bodies where no visceral tissues are available for toxicological examination.

Sample Collection in Suspected Cases of poisoning
(even if, at the time of the necropsy, a toxin is not suspected.)
Take samples and place half of each sample in aluminium foil and the other half in plastic bags or containers (aluminium or plastic interfere with the testing of some toxins). Samples should be stored frozen (if possible) until shipped to a laboratory. If a particular toxin or class of toxins is suspected of being involved in animal morbidity or mortality, one may also collect various other samples, depending on possible routes of exposure viz. Environmental samples., Feed samples., Water samples., Heparinized whole blood (20 ml or more, in blood collection tubes) and urine from live animals.

In most cases, toxicology samples should be stored frozen until tested.

Containers for preservation of materials
Wide-mouthed glass bottles of about 2-litre capacity having airtight stoppers should be used for visceral tissues. These bottles should be numbered and labeled properly which should mention about the details of the case, nature of the contents preserved, place and date of preservation etc., and should bear the signature of the veterinarian.

Preservation of tissues

The tissues are taken into the container and sufficient alcohol is added so that whole tissues are dipped into the solution. Tissues are also preserved in saturated sodium chloride solution. The tissues are taken into the container and sufficient quantity of common salt is added to it. The tissues are immersed well with the salt. Some salt should remain at the bottom of the container and over the tissues. A sample of alcohol or saturated solution of common salt used for preservation must also be sent in separate glass bottles for analysis to exclude the presence of any poison in it.

Goals of Toxicological Investigation

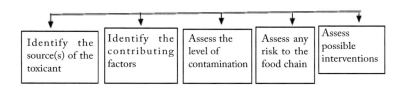

Questions to consider

1. Are storage containers (feed and chemicals) labeled properly?
2. Are feeds stored with chemicals?
3. Is the feed stored appropriately to prevent spoilage?
4. Is there any visible mould on/ in feedstuffs, or do feedstuffs smell mouldy?
5. What are the potential environmental facilities, or home environmental risks? Are there any recent changes in the environment or management?
6. Do the animals have access to trash, building materials, medicines, including those intended for other species, toxic plants, or substances provided by non-caretakers?

Sample handling

- Always wear appropriate personal protective equipment while performing necropsy examinations or collecting potentially hazardous samples. Wearing gloves may also be important in protecting sample integrity.

- Change gloves between samples (this is particularly important to assess differences in contaminant concentrations).

- Collect samples from least contaminated to the most.

- Collect adequate sample amount (250 g +).

- Find out from the laboratory, the appropriate containers in which the specimens are to be sent depending upon the type of analysis required (Glass--not ideal for shipping; may require secondary packaging, Plastic or Aluminum foil)

Properly label with: Date/time, Collector's Name, Species of animal, Tissue.

Collect GI content in the last

- Store the sample properly (Freezing in most cases).
- Some samples require rapid transport to the laboratory.

Materials to be sent to Toxicological Analysis

Suspected Poison for Analysis	Specimen Required	Amount Required	Remarks
Alkaloids	Blood, urine Liver, stomach content, kidney, brain	200 ml 500 g	
All poisons	Bait or feed	500 g	All poisons
Ammonia/Urea	Whole Blood or Serum Urine Rumen contents	5 ml. 5 ml. 100 g.	Frozen or may add 1-2 drops of saturated mercuric chloride*
Arsenic	Liver, Kidney, Whole Blood, Urine, Ingesta Feed	100 g. 15 ml. 50 ml. 100 g. 1-2 kg.	
Barbiturates	Blood Brain, fat	200 ml 500 g	Barbiturates
Chlorinated Hydrocarbons	Cerebrum, Ingesta Fat Liver, Kidney	500 g.	Use only glass containers. Avoid aluminium foil for wrapping specimen.

Collection and Preservation of Biological Specimens

Suspected poison for Analysis	Specimen Required	Amount Required	Remarks
Chronic arsenic	Skin, hair, stomach contents, liver, kidney	500 g	Chronic arsenic
Chronic lead	Bone	500 g	Chronic lead
Copper	Kidney, Liver Serum Whole Blood Feed Faeces	100 g. 2-5 ml. 10 ml. 1-2 kg. 100g.	
Cyanide	Forage Whole Blood Liver	1-2 kg; 10 ml. 100 g.	Rush sample to laboratory. Frozen in airtight container.
Fluoride	Bone, teeth Water Forage Urine, blood	20 g. 100 ml. 100 g. 50 ml.	Ideal sample will be the lesion seen in teeth and bone.
Heavy metals	Urine , blood Stomach contents, liver, kidney	200ml 500g	
Herbicides	Treated weeds Urine Ingesta Liver, Kidney	1-2 kg. 50 ml. 500 g. 100 g.	
Hydrocyanic acid	Blood, Urine Stomach contents, liver, muscle (freeze immediately)	200ml 500g	
Lead, Mercury	Kidney Whole Blood Liver Urine	100g. 10 ml. 100 g. 15 ml.	Heparinized, do not use EDTA
Mycotoxins	Brain, Forages, Liver, Kidney Stomach contents, Blood	100 g. 200ml	Airtight container, plastic bag for dry samples

Suspected poison for Analysis	Specimen Required	Amount Required	Remarks
Nitrate,	Forage Water Body fluids	1-2 kg. 100 ml. 10-20ml.	
Nitrites	Blood Stomach and gut contents (sealed in airtight containers)	200ml 500g	
Oraganoarsenicals	Hair, liver, kidney, stomach contents	500 g	
Organocarbonates	Feed Ingesta Liver Urine	100g. 100 g. 100 g. 50 ml.	
Organophosphates	Blood Brain, fat (for200ml cholinesterase)	500 g	Organophosphates
Oxalates	Urine Kidney, stomach contents	200 ml 500 g	Oxalates
Oxalates	Fresh forage Kidney	100 g. 100 g.	Fixed in formalin
Selenium	Blood Liver, kidney, hair	200 ml 500 g	Selenium
Sodium (Na CI)	Brain Serum CSF Feed Stomach content	100 g. 2 ml. 1 ml. 1-2 kg. 500g	
Strychnine	Urine Stomach contents, liver, kidney	200 ml 500 g	Strychnine
Sulfonamides	Kidney Urine	500 g 200 ml	
Thallium	Urine Kidney, stomach contents	200 ml 500 g	

Suspected poison for Analysis	Specimen Required	Amount Required	Remarks
Warfarin	Stomach contents, liver Blood	500 g 200 ml	
Zinc phosphide	Liver, Kidney, Gastric content	100 g.	

Autolyzed or Decomposed Carcasses

Autolysis of tissues can cause many artifacts in tissues that can superficially be confused with a disease process, but histopathological examination can still distinguish the true lesions if tissues are properly handled. For histopathology, autolyzed or decomposed tissue samples must be handled very gently, held at the edges only, cut with a sharp knife or scalpel and quickly placed in formalin (freezing would further diminish the ability to recognize most changes microscopically). Samples both from changed-looking and more normal appearing parts of an organ will be most useful. Quick cooling or preservation in formalin is recommended. For infectious disease or toxicology testing, freezing or refrigeration of samples should take place as soon as possible.

3 Necropsy Methods

• Postmortem Examination of Wild Animals

Necropsy procedures for certain wild animals are similar to that of domestic animals as indicated below.

Procedure for wild animals	Similar to procedure of domestic animals
Wild ruminants, Deer, Antelope, Camel and Illamas.	Cattle, sheep and goat
Wild horses, Asses, Zebras, Elephants, Hippopotamus and Rhinoceroses	Horses
Swine	Swine
Wild birds	Chicken, Turkey, Ducklings and Geese
Wild carnivores, Wolves, Coyotes, Bears, Lions and Rodents	Cats and dogs.

Wild Mammals

Necropsy should be conducted as soon as possible. If no quick examination is possible, smaller carcasses should be kept cool, after noting all necessary data, but not frozen.

If it is not practical to deliver the entire carcass to a necropsy service collect as many samples as possible. Chill fresh tissues immediately and send (with frozen cold packs) to the Animal Health Diagnostic Center as soon as possible.

A thorough history is important. Give as much information regarding the death of the animal as possible. Attach additional documents, if applicable, such as treatment records or ante-mortem test results.

Take photographs or digital images of carcasses and lesions. Wear gloves and other appropriate personal protective equipment (coveralls or impervious gowns, boots, aprons, masks, goggles or face shields).

General Steps for Necropsy

1. The carcasses of equids (Equidae-Horses Family and other Perissodactyla-odd-toed ungulates (Order)) should be laid on its right side for best handling of the caecum and colon.

2. All ungulates are placed ON THE LEFT SIDE.

3. All carnivores, birds, reptiles, and primates are placed ON THEIR BACK.

4. Primates (all species) and small mammals are dissected while lying on the back.

5. Special devices for laboratory rodents are used---a soft board, or sheet of cardboard covered with plastic foil and four stout pins or hypodermic needles.

6. Very small animals (less than 100 g) can be fixed entirely by opening the body cavity and submerging the entire animal in formalin.

7. If the carcass is presented in a container / wrappings, examine them inside and note the contaminants (e.g. mud, oil) and possible external parasites such as fleas, lice or mites which may have left the host. Presence of maggots within the wrappings may indicate significant carcass decomposition and alert the examiner to search for lesions of fly strike.

8. In order to minimize post-mortem changes in well insulated animals such as sheep, pigs, marine mammals and carnivores the examination should be carried out as soon as possible after death.

9. Weigh the carcass.

10. First, make a thorough visual examination of the carcass. All body orifices should be checked for discharges. Bloody discharges may be indicative of anthrax. Do not move or open the carcass.

11. Caution must be exercised if anthrax is suspected. If anthrax is diagnosed, the carcass should be burned or buried. If possible, the carcass should be protected from scavengers while being checked for anthrax.

12. Note the condition of the skin and pelage. Examine for ulcers, shot holes, tooth marks and external injuries. Turn over the carcass and record any broken limb bones. The scavenging birds would often remove the upper eye and the foxes, jackals or dogs would damage the tail and muzzle.

13. Examine the carcass for external parasites, noting abundance and location. Collect representative specimens and store in 70% alcohol. Pay particular attention to predilection sites, e.g. around the muzzle, eyes, ears and genitalia, on the neck, brisket, tail switch, axillae, groins and hoof clefts.

14. Whenever possible, estimate the time of death of the animal. Putrefaction takes place very rapidly when the ambient temperature is high. Post-mortem changes within the carcass are accelerated when the body temperature is high at the time of death, e.g. in the case of heat stroke, lightning strike, anthrax, tetanus and high ambient temperature and in direct sunlight. Post-mortem gas formation in the alimentary tract, especially in ruminants, should be distinguished from ante-mortem bloating, which itself can be a cause of death. In the latter case, there will be signs of asphyxiation.

15. If the carcass has been exposed to the sun for some time, the superficial musculature beneath the skin may acquire a pale, parboiled appearance ('sunburn').

16. When post-mortem changes are more advanced, the muscles show evidence of autolysis and are often watery and soft. In moderate post-mortem change, the imbibition of blood and bile can be observed. In the former, haemolysed blood leaks out of the blood vessels and is absorbed into the adjacent tissues. This can readily be seen as a dark fringe along the sides of the vessels of the omentum

Necropsy Methods 53

and mesentery. Bile imbibed into the wall of the gall bladder stains the adjacent tissues yellowish green.

17. The condition of the heart may be used as an indicator of the time of death and degree of post-mortem change. Haemolysis of blood can produce a strong, adherent red stain on the lining of the right ventricle. Unclotted blood in the left ventricle indicates that the death was recent and rigor mortis has not yet set in. A disintegrating clot and dark haemolysed blood in the left ventricle is seen when an animal has been dead for 24 hours or more. This indicates that rigor mortis has occurred and that extensive post-mortem changes can be expected.

18. Post-mortem changes are rapid in the brain tissue and considerable autolysis takes place quickly.

19. Make an assessment of the general nutritive condition of an ungulate at the time of death by examining the marrow in one of the femurs. Firm, fatty yellowish or white marrow indicates a well nourished animal. Soft, watery and orange or red marrow indicates that the animal has died of starvation or debilitated by prolonged chronic disease.

I. General External Examination

Record the following

1. Posture (Describe it and take photo).

2. Sex and Estimate the age (neonate/ infant/ juvenile/ adult/ aged/old).

3. Any identifying numbers (tattoo, tags, microchip numbers etc). Retain any physical markers for future reference e.g. radio-transmitters, tags.

4. Any characteristic features (characteristic scars, colour marks, coat colour—albino, leucistic, melanistic).

A. Examination of Skin, Fur (Integument) and Body.

Plan before dissection, what to preserve
(i).Entire carcass (or) (ii). Parts of carcass for non veterinarian purpose depending on the

(a).Importance of destruction of parts for examination. Ex. If diagnosis is important because of danger of epidemics or zoonoses   (or)

(b).Importance of preservation of parts of the specimen (skull, skin etc) or whole specimen for taxonomic purposes.

1. Skin

i.   If the preservation of the skin is necessary,

ii. Do not damage the inner skin surface with hair roots. Otherwise the mounted specimen may loose hairs later)

iii. Avoid cuts in thinly haired parts of the skin. If done, ugly seams might be visible after mounting. It is better to leave the parts of the limbs in the skin. (Vacuum freeze-drying with limbs in the skin may be an adequate method for mounting thinly-haired specimens). Therefore, tests (trying to plug a few hairs in several parts of the body including the abdomen) prior to skin removal should be done in taxonomically valuable specimens.

iv. Do not cut through too much hair. Moistening of the hair may help to prevent it from moving between the scissors).

v.   In carcasses beginning to decompose, the hair may already be loose. Remove the skin very cautiously, otherwise the hairs will fall out leaving a naked skin.

vi. If necropsy is not done immediately, removal of the skin with insulating fur before cooling or freezing may help to cool the carcass down more quickly.

vii. In cases with superficial tumours or lesions supposed to be examined, it may be necessary to leave the parts of skin concerned attached to the body.

viii. Note the general appearance of carcass pelage (fur, spines, whiskers (vibrissae): Wet, muddy, oiled, clean, bloody, moult. (If contaminated, note extent and areas affected).

- If in moult note the extent and take into account the season.

- If there are missing patches of fur...Correlate with wounds, parasites, history of pruritus (itching), entanglement etc.

ix. The condition of the fur may indicate whether the animal was in good condition or in a state of debility prior to death.

Necropsy Methods

x. Check for the presence of external parasites (ectoparasites) - fleas, lice, ticks, maggots etc. Note the species, numbers (accurate or approximate indication) and distribution. Particular attention should be paid to checking the predilection sites for external parasites e.g. armpits (axillae), groin, perineum, hoof clefts, eyes, ears etc. Collect external parasites in 70% ethanol for preservation and further identification.

xi. Part the hair over multiple areas of the body, systematically to examine the skin, for lesions-- bite wounds, papules, macules, pustules, comedones, furuncles etc.

xii. Examine all body orifices, including the anus, vulva, prepuce, cloaca, marsupium (pouch) as appropriate, for evidence of haemorrhage, discharge, parasites, maggots abnormal growths etc.

xiii. Turn over the carcass. Examine both sides of the body.

xiv. Examine the perineum and hindquarters for evidence of faecal scouring suggestive of diarrhoea (scour).

xv. Note the general condition of the fur.

xvi. Record any characteristic features --characteristic scars or colour marks, coat colour and malformations.

xvii. Note if external signs of trauma are present

- Parts of the body missing, changed?

- Broken bones, missing hair, broken teeth etc

- Examine the carcass for wounds.

- Note the age, size, location of the wound, degree of sepsis and characterize them.

- Note for signs of bruising or bleeding into the tissue, indicating that the wounds occurred when the animal was still alive? (Otherwise they might be signs of scavenging).

- Note if blood is visible. If so, clotted/ dried? and find out the cause of wounds.

xviii. Changes in Hair, skin, nails / missing?

2. Take dry weight of the carcass. If wet, it should be noted for future reference to allow recognition of potential bias.

3. Take biometric measurements using graduated calipers (tibial length, crown-rump length).

4. Assess the Nutritional state / condition of the body using a combination of subjective (using indices based on muscle bulk, amount of subcutaneous and visceral fat deposit) and objective measurements (by creating indices of parametric measurement to body mass). Dehydrated? Take radiographs to detect fractures, radio dense foreign bodies, air gun or shot gun pellets, metabolic bone disease.

5. Note whether the carcass is fresh or decomposed; whether it has been refrigerated or frozen; also whether it is intact or scavenged, and if scavenged to what degree. Note Rigor mortis, other signs of death, and externally visible signs of decomposition?

6. Estimate time of death where unknown. Autolysis will be accelerated where the temperature of the animal was increased at the time of death e.g. heat stroke, lightning strike.

7. Differentiate gas production in the gastro-intestinal tract post-mortem with pre-mortem 'bloat'

Sample Collection

i.   Hair - For DNA analysis or for a taxonomical reference.

ii.   Parasites in 70% ethanol--For preservation and further identification.

iii. Skin samples--Microscopic (or hand lens) examination of scurf, scale, hair pluckings, deep skin scrapes, Sellotape (clear sticky tap) impression smears, Woods lamp (UV) examination for dermatophytes, swab or sample for microbiology (bacteriology, mycology), lesion sample for histopathology, lesion sample for virology.

iv.   Tissue samples, swabs and impression smears from externally visible abnormal areas (pus, sample from areas with abscesses, samples from the edge of affected (abnormal) areas most likely to contain organisms) with sterile instruments and stored in sterile container-for culture

v.   Skin piece in 10% formalin ...For histopathology.

Appropriate chain of custody and sample labeling should be followed when dealing with legal cases.

B. Examination of Head and Neck.

- Examine the eyes, ears, nostrils (external nares) and mouth for the presence of haemorrhage, discharge, external parasites, maggots, abnormal growths, foreign bodies, other lesions- erosions, ulcers etc. and condition of teeth . In deaths due to acute plant poisoning animals may still have remnants of the plant material within their mouths. Haemorrhage in the skin around the eyes may be a sign of a skull base fracture.

- Examine the eyes carefully for evidence of opacity or long standing injury (free-ranging predators may lose condition and potentially starve if they rely on acute vision to catch prey).

- Examine ears for blood, pus and parasites.

- Note the colour of mucous membrane of the mouth and conjunctiva for evidence of jaundice or pallor (suggestive of blood loss).

- Examine the condition of the lips, oral mucosa, soft / hard palate and tongue for ulcers, growths or developmental abnormalities (e.g. cleft palate).

- Cut through the oral commissures and examine the back of the oral cavity.

- Continue the incision and disarticulate the lower jaw. Inspect the nasopharyngeal area, tonsils, salivary glands and retropharyngeal and parotid lymph nodes.

- Examine the teeth for evidence of tooth loss, abnormal or excessive wear (attrition), gingivitis, periodontal disease, tartar accumulation etc. Record the dental formula.

- Examine the teeth for normal or irregular wear and the gums and alveoli for abscesses. Open the nasopharynx and examine for bot fly larvae in the retropharyngeal pouches. Incise and examine the tonsils and retropharyngeal lymph nodes for abscesses and tubercular lesions. Examine the tongue for ulcers and for cycticerci and macrosarcocysts in the musculature. Locate, incise and examine the parotid lymph nodes.

- Remove the skin from the skull and the temporal muscles.

- Examine the subcutaneous tissues and skull table for evidence of trauma, bruising, etc.

58 Wildlife Necropsy and Forensics

- Section the base of the auricular (ear) cartilages and examine the contents for evidence of discharge, inflammation, parasites etc.

- Section the skull to examine the sinuses, turbinates or tympanic bullae if required.

- Cut through the head in front of the eyes with the help of a saw and examine the sinuses and turbinate bones for infection, bot fly larvae (sheep and antelopes), pentastomes (carnivores) and nematodes (mustelids). Saw the cranium lengthwise and examine the brain after removal Or carefully saw around the skull cap and lift it off. Make a careful examination to look for evidence of haemorrhages, bruises, nematodes, tapeworm cysts, bot fly larvae, abscesses or inflammation of the brain tissue or meninges.

Examination of Head: Ruminants

The head with the salivary glands intact is severed through atlanto-occipital joint. In ruminants cut the skin and muscles at the angles of the mouth and disarticulate the lower jaw. Examine the teeth for normal or irregular wear and the gums and alveoli for abscesses. Cut the hyoid bones and remove the floor of the mouth, including the tongue.

Open the nasopharynx and check for bot fly larvae in the retropharyngeal pouches. The tonsils and retropharyngeal lymph nodes should be incised and examined for abscesses and tuberculous lesions.

Examine the tongue epithelium for ulcers on the surface and cysticerci and macrosarcocysts in the musculature. Locate the parotid lymph nodes, incise and examine.

Remove the skin over the head and the temporal muscles. Severe the ears and check the aural canals for spinose ear ticks (Otobius megnini), nymphal Rhipicephalus spp. ticks and ear mites. These may be embedded in masses of wax.

Use a saw to cut through the head in front of the eyes, examine the sinuses and turbinate bones for infection, fly larvae etc. Remove the brain and examine by sawing the cranium length-wise or by carefully sawing around the cap and lifting it off. Examine carefully for evidence of haemorrhaging, bruising, nematodes, tapeworm cysts, bot fly larvae, abscesses or inflammation of the brain tissue or meninges.

Necropsy Methods

Sample Collection

i.   Collect tissue samples, swabs for culture and impression smears while examining the tissues first.

ii. Collect tissue (no more than 10 mm thick, and thinner if congested) in at least ten times the volume of formalin fixative as the tissue for histopathology.

iii. Collect external parasites in 70% ethanol for preservation and identification as required.

iv. Eyes may be removed with care and fixed in Bouin's or Zenker's solution.

Appropriate chain of custody and sample labeling should be followed when dealing with legal cases.

C. Musculo-skeletal system - External Examination.

Examine the limbs for evidence of fracture, dislocation, swelling, crepitus, deformity, lacerations, wounds (including snare wounds) etc.(Use the other limb as a comparison for reference to help identify pathology).

Palpate the muscle bulk over all limbs (indicator of body condition). (Muscle wasting (atrophy) affecting a single limb, or group of muscles within a single limb, may indicate recent disuse, possibly due to a long-standing injury affecting the limb). Manipulate each limb and joint through its range of movement (if rigor mortis has not set in) comparing each side with one another, and systematically working from the bottom (distal) to the top (proximal) of the limb. Abnormal increased or restricted range of movement at a joint should direct further internal joint examination. Examine radiographically if an abnormality is detected on palpation.

Sample Collection

Collect tissue for histopathology
Appropriate chain of custody and sample labeling should be followed when dealing with legal cases.

II. General Internal Examination

1. Wet the fur thoroughly with soapy water or disinfectant in water (to minimize debris which may contaminate the internal organs).

2. Lay the carcass on its back or side (Small carcasses may be kept in place using metal pins secured into a wooden board).

3. Cut along the inner aspects of both hind limbs over the groin deep into the hip joints, severing their articular connections, until both limbs can then be placed out to the side.

4. Cut through the inner aspects of both forelimbs over the armpits (axillae), disarticulating the shoulders by cutting through the muscular attachments inside the shoulder blade (scapulae).

5. All four limbs, still attached to the torso by soft tissue, should then fall to the side, providing a stable base for further examination.

6. Examine the reproductive organs.

a) In males examine Testes

i.   whether inguinal or scrotal

ii. Size

iii. Pattern of skin of the scrotum and

iv. Other features—secretions etc

Examine the prepuce and penis. Detach the penis along with its prepuce and draw backwards up to ischial arch. Expose the testicles by cutting the scrotum. Incise the tunica vaginalis and draw out the testicles with the help of scissors. Cut the tunica vaginalis and separate the spermatic cord and leave it as such hanging on either side of the abdomen

b) In female examine

the external genitalia, the vulva. Record external signs of reproductive status.

Condition of the vagina
(a) Open / closed / sealed by a skin.?

Necropsy Methods 61

i. A sealed vaginal cleft may occur in juvenile females and in seasonal breeders at certain times of the year, with the vaginal opening closed by a membrane which may look like normal skin with vaginal opening absent.

ii. Signs of estrus such as swollen and reddened rims of vaginal opening?.

iii. Signs of pregnancy?

iv. Examine the mammary glands, where present and note their state of development and activity-- signs of lactation? (Can milk be squeezed from the teats?) Nipples looking used?. **Remove the mammary glands.** Examine cut surface of the gland and palpate for evidence of scarring, mastitis, etc. Offspring: babies / infants clinging to the adult found, babies / infants found parked in the proximity

v. In neonates, examine the umbilicus.

7. Cut open the skin from the mandible to the pubic area in the ventral midline (from chin to the pubic bone), taking care not to damage the underlying tissues.

8. Reflect the skin from the neck, chest and abdomen and examine the subcutaneous tissue noting the amount of subcutaneous fat, colour of subcutaneous tissues, bruising, haemorrhage or other lesions. Remove the skin to allow complete examination of the subcutaneous layers. An examination of the inner surface of the removed skin may show wounds hidden under the hair or effusions of blood. In addition muscles and fat can be examined. Check for evidence of bruises, wounds, parasites, ballistic wounds, snake bites, burns, etc.

9. Make an incision on the abdominal wall along the mid line (in the linia alba), from a central point working caudally (backwards)— towards the pubic bones-pubic symphysis and cranially (forwards) towards the base of the rib cage and diaphragm (xiphoid cartilage), guided by two fingers taking care not to damage underlying tissues(in case, there is tense wall due to bloat etc,). Take care to avoid opening the gastro-intestinal tract inadvertently when incising the abdominal wall. Accumulation of gas in the gastrointestinal tract occurs postmortem, stretching the intestinal loops and making accidental damage early in the examination . Contamination of the abdominal cavity with gastro-intestinal contents may obscure visualization of gross pathology and interfere with bacteriological examination. Reflect the abdominal wall along the costal arch (just below sternum) to the level of the vertebral column on either side and in females along the border of the pubic bones as well. In males the incision is continued towards inguinal canal.

10. Examine the integrity of the diaphragm, checking for hernia, rupture, etc. If no skeletal preservation for a collection is necessary, the thorax may be opened.

11. Cut the ribs on both sides through the costo-chondral junctions -weakest point at the junction between cartilage and bone (in small mammals with bone scissors, for larger species a saw would be necessary). Remove the sternum with attached parts of rib cage. Alternatively cut along the region where the ends of rib bones are connected to cartilage, where cutting is easier and the bony parts of skeleton are less damaged. Lift the sternum upwards and remove the rib cage by dissecting free remaining attachments at the thoracic inlet, cutting through the clavicles as necessary. The rib cage can be laid over to act as a tray upon which organs can be placed for examination.

12. Saw the pubic bones to allow full examination of the pelvic organs.

After the body cavities are opened

13. Examine the chest (pleural) and abdominal (peritoneal) cavities for evidence of abnormal locations or size of organs, haemorrhage, exudate, transudate and effusions. Note the quantity, color, smell, viscosity , turbidity and volume of the fluids. Collect samples of the fluid for further testing as required. A small amount of pale yellow fluid in these two sites is normal. Note the appearance (smooth, moist and glistening or dull and granular with adhesions) of the lining membranes of the chest (pleura)and abdominal cavities (peritoneum). Examine pleural and peritoneal cavities for evidence of adhesions (attachments of organs to the body cavity and determine if these attachments are easy to break), parasitic cysts, tumours, inflammation, infection etc.

14. Examine the internal organs in situ. Assess the location of all organs (to determine if any organs are displaced) before organs are removed, study their relative layout, size and colour. Note displacement of any organ, or abnormal attachment of any organ to the body cavity if any.

15. Comment on the amounts of abdominal and mesenteric fat deposits (indicates the general nutritional condition of the animal).

- At this time, collect a sterile blood sample for culture from the heart (the right atrium is the best location), and additional blood in a vial to obtain serum for serological tests.

- Take sterile samples of other organ for culture before organs are handled.

Necropsy Methods

- After recording the general condition of the animal, individual organs can be removed, examined (Examine and palpate each organ for evidence of pathology; consider colour, consistency and shape in every case. Organs should be palpated methodically by carefully feeling the tissue between the thumb and forefinger), and sampled in a systematic manner. Record any abnormal findings (lesions) and take photographs (provide the best documentation for records).

- Bile staining of tissues adjacent to the gall bladder can be seen in carcasses as post-mortem change.

16. Next, examine the spleen and inguinal lymph nodes. In large ruminants, especially cervids, deflect the rumen to locate the spleen.

If the spleen appears grossly enlarged and swollen and the blood is black and tarry, anthrax can be suspected (a blood smear should be made, stained and examined before proceeding any further with the autopsy). If the stained blood smear is negative for anthrax, continue the postmortem examination.

General inflammation of the spleen and viscera, with or without abscesses in the lymph nodes, may indicate an infectious condition.

The amount of omental fat, kidney and heart fat should be recorded. The fat should be weighed after dissecting it out, if necessary.

Sample Collection

i. Use the removed ribs for examination of the stability (ossification) of bones and bone marrow examination..

ii. Collect tissue samples, swabs for culture and impression smears from all abnormal looking areas, (pus, from areas with abscesses or nodules). Organisms are present at the edges of affected (abnormal) areas.

iii. Collect tissues for histopathology (no more than 10mm thick, and thinner if congested) in at least ten times the volume of formalin fixative as the tissue.

iv. Retain samples of the fluid for further testing as required.

v. Collect tissue sample from diaphragm to detect Trichinella infection (Trichinellidae-Family) in carnivores, and wild pig species (Suidae-Pigs (Family). Collect tissue samples for toxicology- wrap in foil (if the suspected toxin is organic), plastic wrap (if the suspected toxin is metal) or glass (if suspected toxin is unknown). Fat samples are important for toxicological investigation (e.g. organochlorines).

Appropriate chain of custody and sample labeling should be followed when dealing with legal cases.

System-wise Examination

A. Cardio-vascular system.

i.   Examine the heart and great vessels *in situ* looking at their relative layout (e.g. patent ductus arteriosus (PDA), persistent right aortic arch), size, shape and colour.

ii. Remove the organs of the chest by cutting the trachea and great vessels as far cranial (forward) as possible. Applying gentle traction, carefully dissect back the 'pluck' containing the thymus (large if young, remnant may be identified in adult), heart and lungs. Alternatively, dissect along the inside of the blades of the lower jaw, free the tongue, larynx, trachea and oesophagus and remove together with the 'pluck'.

iii. Examine the 'pluck' as a whole and then dissect each organ free for individual examination.

iv. Removal and examination of the heart: If no changes in the blood vessels have been noticed during opening and first examination, cut the major blood vessels close to the lungs. Cut the aorta and posterior vena cava close to the diaphragm, remaining connected with the heart for removal with the heart. Examination of the heart should include thickness of walls (right, left ventricle, septum), the inner surface (papillary muscles), the entire septum muscle (cut), and collection of samples (several sections of ventricle muscle, particularly close to the papillary muscles and rhythmic center).

v.   Dissect out the heart, leaving a portion of the major blood vessels intact.

vi. Examine the surface of the pericardium, incise the pericardial sac and remove from the surface of the heart. Note the volume, colour, turbidity and viscosity of any fluid present. Note the presence of adhesions between the pericardium and the surface of the heart.

Necropsy Methods

vii. Examine the overall shape and symmetry of the heart and its epicardial (outer) surface. Record the weight. Take following measurements: circumference of the heart in the atrioventricular groove (Sulcus coronarius) and height of left and right ventricle (maximal length and width).

viii. Systematically dissect the heart using scissors to follow the path of blood through the heart. First examine the right side chambers, followed by the larger and thicker walled left side chambers. **Opening of the heart:** This is done usually by following the normal path of blood flow, beginning in the right atrium. A u-shaped incision is made through the right atrio-ventricular valve, following the inter-ventricular septum to the apex, then back to the base of the heart through pulmonary valve. Lift the ventricular wall flap, cut loose this way, to expose the ventricle and valves. Open the left side of the heart in a similar way, the incision beginning in the left atrium, passing through the atrio-ventricular valve to the apex and back through the aortic valve into the aorta. Evaluate the heart muscle and relative thickness and size of the heart chambers (look for dilatation and hypertrophy). Examine the endocardium (inner lining) of the heart and the valves for evidence of irregularities, plaques, thrombus formation, calcification, etc. Examine the surface of the heart for any cysticerci or pale areas. In recently dead animals, the right heart usually contains plasma clots (yellow / tan material), unclotted blood, or clotted blood. Postmortem blood clots inside the heart look smooth whereas a thrombus looks rough and attached. A yellowish inner surface of the aorta is a sign of jaundice. Note the thickness of walls (right, left ventricle, septum). Examine the inner surface (papillary muscles) and the entire septum muscle. Incise the ventricular muscle at several places for lesions.

ix. The appearance of blood within the heart chambers can be used to help estimate the time since death.

x. Unclotted blood in the left ventricle is suggestive of recent death and examination pre rigor mortis.

xi. Disintegrating blood clots within the left ventricle suggest that the animal has been dead for in excess of 24 hours and has passed through the period of rigor mortis.

xii. Examine the heart valves for granular lesions and the aorta and carotid arteries for filarid worms.

xiii. Cut the myocardium at a number of places and check for muscular cysticerci (which are also sometimes found in the coronary fat), the presence of any discolourations, parasitic cysts, growths, infarcts, scars, etc.

Heart: At death, the heart will be in diastole. The rigor mortis develops in the left ventricle within an hour after death due to which it becomes empty but the right ventricle will be about half full. The heart will be flabby. If at this time if the heart is flabby and the left ventricle is filled with coagulated blood, one must suspect for myocardial degeneration. If the heart is filled with uncoagulated blood one must suspect for asphyxia and haemolysis (though rigor mortis has occurred, factors needed for post-mortem blood coagulation are not available in sufficient quantities). If any patchy discolouration of the heart is observed one may suspect for diseases of blood or poisoning. Dilatation usually occurs on the right side. Hypertrophy on the right side will increase the width of the heart at the base, while the hypertrophy of the left will increase the length of the heart. In bilateral hypertrophy, the heart will assume a round shape. The blood clots rapidly after death from acute inflammation / fever.

Blood remains fluidy in cyanide, alcohol, chloral hydras, copper, benzene, Sweet clover poisoning, snake venom, Anthrax (Tarry), heat stroke (left ventricle empty), Phosphorus poisoning, asphyxia, chronic lead poisoning, Traumatic pericarditis, septicaemia and toxaemia.

Coagulated blood is seen in Septicaemia,
Partly coagulated blood is noticed in Mercury poisoning (dark red and coagulates slowly), burns, Urea poisoning and Electrocution
Chocolate brown colour blood is seen in Nitrate / Nitrite poisoning
Cherry red (slightly brown) colour blood is seen in Sodium chloride poisoning
Bright red or pink colour blood is noticed in cyanide / Hydrocyanic acid poisoning
Dark brown colour blood is seen in Carbon dioxide poisoning and
Cherry red colour blood is seen in Carbon monoxide poisoning.

Petechial haemorrhages are noticed in Septicaemia, Asphyxia and toxaemia. In septicaemia, toxaemia and asphyxia, petechial haemorrhages over the serous membranes, epicardium and endocardium will be noticed and the blood will be dark and imperfectly clotted.

Note the colour and amount of epicardial fat. Note the relative size and position of the heart, presence of pale, mottled, discoloured areas or scarred areas in the myocardium

Necropsy Methods

thickness of the walls of ventricles, thickness of the valves, congenital defects etc.

Sample Collection

i. Collect tissue samples, swabs for culture and impression smears as tissues are examined.

ii. Collect tissue samples for histopathology

iii. Collect blood samples from the heart chambers for bacteriological examination before the heart is handled or incised to reduce contamination. If heart blood swabs are to be collected, heat the spatula and singe the heart. Heat the tip of the scissors or scalpel, open the heart over the singed area keeping the organ very close to the spirit lamp. Take out the swab from the test tube, dip into the blood in the chamber of the heart. Alternatively the blood can be collected from the heart with the help of a Pasteur pippet. Samples of clotted blood are taken from the left ventricle for serology. Blood clots are centrifuged to collect serum samples for further testing.

Appropriate chain of custody and sample labeling should be followed when dealing with legal cases.

B. Respiratory system

i. Examine the lungs, trachea (windpipe) and major bronchi (airways) in situ; study their relative layout, size, shape and colour.

ii. Visually examine and palpate the lungs for evidence of pathology. Note the colour, consistency and shape.

iii. Using sharp scissors, cut along whole length of trachea and into the major bronchi, noting presence of fluid, blood, fungal plaques, necrotic lesions, parasites, bot fly larvae, foreign bodies, etc.

iv. Cut through the lungs in serial sections, noting the presence of any water (edema), froth, blood, fungal infection, abscesses, tumours, parasites, tuberculous lesions, etc. Squeeze the lung tissue and observe for parasites within the airways.

v. Placing a cut section of lung tissue within water may help evaluate congestion, consolidation, emphysema etc.

vi. Examine the tracheo-bronchial lymph nodes visually and by palpation. Cut serial section for evidence of enlargement, focal lesions, abscesses, tumour metastases, tuberculous granulomata etc.

vii. Record the weight and organ dimensions (maximal length and width).

Sample Collection

Take tissue/swab for culture, impression smears from cut surface, tissue for histopathology and toxicology as appropriate.
Appropriate chain of custody and sample labeling should be followed when dealing with legal cases.

C. Examine the thyroid and parathyroid glands, the endocrine pancreas and the adrenal glands, first in situ, then dissect free and examine in detail. Visually examine and palpate the organs for evidence of pathology; consider colour, consistency and shape. Record the weight and organ dimensions (maximal length and width).

In carcasses of rare species, preservation of organs in situ for anatomical studies may be useful, with only small samples taken out for examination by causing as little disturbance as possible. If a necropsy is done only for detection of health problems, enough tissue for examinations including for bacteriology should be taken. Samples including abnormal areas with surrounding normal areas, not thicker than 1 cm (for good fixation), but long and wide enough to represent different areas of a tissue and possible abnormalities are collected. In small animals, entire organs instead of samples may be collected.

Sample Collection

Take tissue/swab for culture, impression smears from cut surface, tissue for histopathology and toxicology as appropriate.
Appropriate chain of custody and sample labeling should be followed when dealing with legal cases.

D. Liver

i. Ensure knowledge of the normal shape and number of liver lobes in the species under examination since significant variation exists between taxa.

Necropsy Methods

ii. First examine the liver in situ; study its layout, size, shape and colour and relation to other abdominal organs.

iii. Dissect the liver lobes free of their attachments and remove.

iv. Visually examine and palpate the liver for evidence of pathology. Note the colour, consistency and shape. Note whether any cysts present under the capsule. Old cysts may be calcified.

v. Note any haemorrhage on the liver surface or free in the body cavity; whether the capsular surface is intact or split; the presence of pale areas or other discolouration; whether lesions are flush with the surface, protruding or shrunken; whether edges of liver lobes are sharp (normal) or rounded (enlarged).

vi. Record the Weight and organ dimensions (maximal length and width).

vii. Cut through the liver lobes in serial sections, noting the presence of any colour variations, size and shape of any lesions (pale areas, haemorrhages, abscesses, growths, cirrhosis, etc.).

viii. Note the size of the gall bladder. Estimate the volume of bile present. Note the colour and consistency. Note the degree of post-mortem local bile staining of tissue. Carefully express the gall bladder to determine whether any blockages may be present. Dissect open the gall bladder and look for any material within the bile e.g. stones, parasites. Examine for liver flukes (some species are very small), liver tapeworms and liver nematodes. The liver should be sliced at a number of places in order to detect any abscesses or cysts, which may be present deep in the substance.

Sample Collection

i. Take tissue/swab for culture, impression smears from cut surface, tissue for histopathology and toxicology as appropriate.

ii. Samples for toxicology should be either wrapped in foil (if the suspected toxin is organic), plastic wrap (if the suspected toxin is metal) or glass (if suspected toxin is unknown). Liver samples are important for toxicological investigation (e.g. organochlorines).

Appropriate chain of custody and sample labeling should be followed when dealing with legal cases.

E. Spleen

1. Ensure knowledge of the normal shape of the spleen in the species under examination since significant variation exists between taxa.
2. Examination of the spleen is recommended early in the post mortem after the abdomen is cut open. If the spleen is enlarged, with a black and tarry bloody appearance, testing for anthrax infection should be performed before progressing further.
3. First examine the spleen in situ, study its layout, size, shape and colour and relation to other abdominal organs.
4. Dissect the spleen free of its attachments and remove recording the presence of any parasitic cysts. Remove the omentum carefully and completely and weigh if data on fat depots are required. The visceral peritoneum should be examined for signs of inflammation and adhesions.
5. Visually examine and palpate the spleen for evidence of pathology; consider colour, consistency and shape.
6. Note any haemorrhage on the surface of the spleen or free in the body cavity; whether the capsule surface is intact or split; the presence of pale areas or other discolouration; whether lesions are flush with the surface, protruding or shrunken; whether edges of the spleen are sharp (normal) or rounded (enlarged).
7. Weigh and record organ dimensions (maximal length and width). Splenic dimensions can be artifactually increased if the animal was euthanized with barbiturates.
8. Cut through the spleen in serial sections, noting the presence of any colour variations, size and shape of any lesions (pale areas, haemorrhages, abscesses, growths, etc.).

Sample collection

Take tissue/swab for culture, impression smears from cut surface, tissue for histopathology and toxicology as appropriate.

Appropriate chain of custody and sample labeling should be followed when dealing with legal cases.

F. Kidneys/Uro-Genital System

1. Ensure knowledge of the normal shape of the kidneys in the species under examination, since significant variation exists between taxa.

2. Ensure knowledge of the normal shape of the uterine tract in the species under examination, since significant variation exists between taxa. e.g. uterus simplex, uterus bicornuate, uterus duplex.

3. First examine the urinary system and then the reproductive system in situ. Study their layouts, size, shape and colour and relation to other abdominal organs.

4. Trace the flow of urine through the urinary system in situ; follow the ureters from the kidney to the bladder and the urethra as possible. Check for developmental abnormalities e.g. ectopic ureter.

5. Dissect the kidneys free of their attachments and remove.

6. Note the amount of fat deposits present over the kidneys.

7. Visually examine and palpate the kidneys for evidence of pathology; Note the colour, consistency (firm, soft, pulpy) and shape.

8. Note whether the capsule is adherent or easily detached from the kidney surface.

9. Examine the cut surface of serial sections of both kidneys in two planes (longitudinal and transverse). Study the distinction between the cortex and medulla. Note the presence of any abscesses in the kidney cortex, cysts, stones (uroliths), parasites, or etc. Record the consistency of the kidney substance (firm or soft and pulpy).

10. Record the Weight and organ dimensions (maximal length and width). Remove the kidney fat and weigh if necessary. Open the ureters and note the condition of the linings.

11. Open the bladder, examine its mucosa and contents for abnormalities.

12. Examine the external genitalia including the vulva, prepuce, penis, and testes.

13. **Female reproductive tract:**

    Ovaries: Note the size, whether ovaries are active (follicles present and growing, corpora lutea) or inactive, and examine serial cut sections of both ovaries. Note the appearance of oviducts Dissect the tract. Open from the vulva through to the tips of the uterine horns. Note the presence and number of fetuses (estimate the stage of gestation, mummification), implantation scars (in rodents and carnivores), presence of retained placentae, evidence of inflammation and infection or growths.

14. Male reproductive tract:

Examine the contents of the inguinal canal. Check for hernia.
Testes: Note the size, shape, colour, growths, active or regressed. Examine serial cut sections of both testes.

Sample Collection

i.   Take tissue/swab for culture, impression smears from cut surface, tissue for histopathology and toxicology.

ii. Samples for toxicology should be either wrapped in foil (if the suspected toxin is organic), plastic wrap (if the suspected toxin is metal) or glass (if suspected toxin is unknown). Kidney samples are important for toxicological investigation.

Appropriate chain of custody and sample labeling should be followed when dealing with legal cases.

G. Gastro-Intestinal System

1.   Examine the gastro-intestinal tract in situ. Note any distension, abnormal layout (twisting or volvulus), discolouration, haemorrhage or lesions on the serosal surface. Barbiturate crystals may be seen on the surface of the intestines if the mammal has been euthanized by intraperitoneal injection.

2.   Ensure knowledge of the anatomy of the gastro-intestinal system and the primary site of microbial fermentation (distinguish between simple stomached, foregut and hindgut fermenters).

3.   Remove the entire digestive tract. First tie off the oesophagus above or at the level of the diaphragm and the rectum as low as possible near the anus. Cut above and below these ties respectively. Carefully dissect the whole gastro-intestinal tract free of its mesenteric attachments, and lay the tract out as one continuous strip.

Necropsy Methods

4. Collect samples for bacteriological examination before further handling of the tract to limit contamination. Sear the surface of the area to be sampled with hot metal and introduce a sterile pipette to take the sample material.

5. Use clips or ties to isolate the contents of different sections of the gastro-intestinal tract. Tie at the pylorus to isolate the gastric contents, and at the ileo-caecal valve to isolate the small intestinal from the large intestinal contents. Collection and examination of the gut contents from each section of the gastro-intestinal tract is particularly useful for parasitological examination. Quantitative estimates of numbers of parasites present can be made with further techniques.

6. Systematically examine the entire gastro-intestinal tract, usually from oesophagus to rectum, including the pancreas. Use scissors or an enterotome to open up the tract. Note ulcers, inflammation, haemorrhage, tumours if present.

7. Note the consistency, colour and volume of the gut contents. Note whether the mammal has eaten recently and if possible, what, it was feeding on. Note any foreign bodies or parasites present. Note whether formed faecal pellets are present in the rectum.

8. Examine the mesenteric lymph nodes visually, by palpation and cut serial sections for evidence of enlargement focal lesions, abscesses, tumour metastases, granulomata etc.

9. In the equidae, pay particular attention to examine the blood vessels (abdominal aorta, anterior mesenteric artery) to the large colon and caecum for the presence of thrombi, aneurysms or scars caused by parasites. *This is accomplished by opening the abdominal aorta longitudinally and tracing and opening the anterior mesenteric artery and its branches.*

The stomach, trachea and soft palate of equids should be examined for bot fly larvae.

Sample Collection

i. Collect tissue samples, swabs for culture and impression smears and tissue for histopathology.

ii. Examine the fresh samples of contents and mucosal scrapings of duodenum, ileum and caeca microscopically for parasites, parasitic ova and coccidial oocysts.

iii. Samples of gastric contents should be taken when investigating a potential poisoning case. The examiner must decide whether the entire content of the intestinal tract should be preserved for food analysis or whether other examinations may be more important. Before preserving the intestine or longitudinally opening it for examination, sections from different areas may be closed with thread and removed for later laboratory examination.

Preservation of samples from the intestine should include multiple sections from different areas, about 3 cm² of duodenum, Jejunum, Ileum (section close to caecum), Caecum, and a piece of rectum with content for test for parasites and pancreatic insufficiency. Examine the Omentum for parasitic cysts.

Examine the mediastinal lymph nodes visually, by palpation and on cut section for evidence of enlargement, focal lesions, abscesses, tumour metastases, granulomata etc.

Appropriate chain of custody and sample labeling should be followed when dealing with legal cases

H. Lymph Nodes

1. Examine the superficial and deep lymph nodes from both sides of the body visually, by palpation and on cut section for evidence of enlargement, focal lesions, abscesses, tumour metastases, granulomata etc.

2. Record the weight and organ dimensions (maximal length and width).

Sample Collection

i. Collect tissue/swab for culture, impression smears from cut surface and tissue for histopathology

Appropriate chain of custody and sample labeling should be followed when dealing with legal cases.

I. Musculo-Skeletal System – Internal

1. Incise and examine the major limb joints in turn, comparing both limbs for reference in each case. Note the volume, colour, viscosity and turbidity of

Necropsy Methods   75

synovial fluid present. Examine joint surfaces and tendon sheaths for evidence of discolouration, irregularity, degeneration, joint mice, etc. The joints are examined by opening from the medial side commencing distally on the limbs.

2. Check the state of mineralization of the major limb bones by seeing how difficult they are to bend or break.

3. Incise and examine the leg muscles over multiple areas of the body. Note the muscle colour (e.g. pale areas with white muscle disease and capture myopathy, dark appearance with clostridial infection) or for any haemorrhage.

4. Examination of the bone marrow can be used during the post mortem examination of ungulates to judge the nutritional status. A femur of a well nourished animal will have yellow/ white firm and fatty marrow in comparison with orange/red, watery and soft marrow in animals suffering from starvation or debility due to prolonged chronic disease.

Sample Collection

i. Take tissue/swab for culture, impression smears from cut surface and tissue for histopathology. Appropriate chain of custody and sample labeling should be followed when dealing with legal cases.

Completely remove the skin and examine the outer surface of the body of the animal for evidence of bruising or wounding, subcutaneous nematodes, warble fly larvae, bullet wounds and snake bites. Animals struck by lightning sometimes show evidence of burns on the subcutaneous tissues. Pale areas in the musculature may indicate 'sun burn' or 'white muscle disease' (myopathy). The muscles may be discoloured by clostridial infections ('black quarter'). Joints may be swollen and affected by septic arthritis.

J. Nervous System

1. Autolysis of the brain occurs particularly rapidly. Priority should be given to removal of the brain for fixation if required.

2. First disarticulate the skull at the level of the atlanto- occipital joint and remove all overlying musculature.

3. Use an appropriate saw to cut through the cranium. The techniques may vary

with the species under examination. Make a transverse cut across the width of the skull just behind the orbits (eyes). Make two cuts joining with the outer limits of the first cut, towards the foramen magnum. Alternatively cut around the skull cap. A medical vice used to stabilize the skull in position may be useful if available.

4. Carefully lift off the top of the skull using blunt elevation.

5. Examine the brain, meninges and the cerebral vessels in situ. Note lesions if any i.e. for evidence of local haemorrhage, bruising, swelling, redness or pale colouration, parasites abscesses, growths or inflammation, etc.

6. Remove the brain from the skull through careful elevation, cutting the cranial nerves in turn.

7. Weigh and record organ dimensions (maximal length and width).

8. Make incision on both the cerebral hemispheres in the middle to expose lateral ventricles. Examine the cut surface of serial sections of the brain, noting the distribution of any lesions found. The distribution of focal lesions found on post mortem examination should be related to the clinical history where neurological signs were noted.

9. Make incision around sella tercica and scoop the pituitary.

10. Cut open the head longitudinally and examine the sinuses, turbinates, eyes, nasal chambers, mouth and pharynx.

11. Dissection for examination of the spinal cord can be time consuming. Dissect away the overlying musculature, carefully trim the vertebral arches to allow visualization of the spinal cord. The spinal cord is exposed by sawing through the arch of the vertebrae. It can also be exposed by cutting through edge of the spinal canal, slightly off the midline.

Sample Collection

i. Collect tissue/swab for culture, impression smears from cut surface and tissue for histopathology.

ii. Fixation of the entire brain may be required. Remove the top of the skull and place the skull and brain in fixative overnight. The brain will then become increasingly firm and can be easily removed from the skull without damage.

Necropsy Methods

iii. Brain samples are important for toxicological investigation (e.g. Organo-chlorines, Organophosphates and carbomates). Samples for toxicology should be either wrapped in foil (if the suspected toxin is organic), plastic wrap (if the suspected toxin is metal) or glass (if suspected toxin is unknown).

Appropriate chain of custody and sample labeling should be followed when dealing with legal cases.

Opening of Thoracic Cavity: In Small Ruminants

The chest is opened by extending the abdominal incision through costal cartilage using a knife in the case of young animal and a saw for an old animal. Hold the cut edge of the rib cage and lift sharply upwards breaking the ribs near the articulation with the vertebrae (In large mammals, an axe or large saw must be used to cut through the ribs at the vertebral level.).

Removal of Neck and Thoracic Organs : In Large Ruminants

Cut the diaphragm at its insertion at xiphoid cartilage and along the costal attachments on either side and at sub-lumbar insertion and remove it. Do not damage the pericardial sac. Free the pericardial sac from the sternum. Separate the oesophagus and trachea in the neck region and divide them in the middle. Make a short transverse incision between the 2 cartilaginous rings of trachea and insert fingers for grip and pull the free end of trachea along with oesophagus towards thoracic inlet. Push the trachea into the thoracic cavity and pull backwards along with the thoracic organs separating them from their attachments through the mediastinum and brachial vessels. Remove the thoracic aorta with these organs and cut through the lumbar region, leaving the abdominal aorta.

Removal of structure of oral cavity : Incise along the medial border of the mandible on either side as far as the symphysis. Draw out the tongue and push backwards simultaneously extending the incision backwards. Incise and cut the soft palate from each side forward and medially to meet in the middle line forming a triangular area. Divide the hyoid bone on either side by placing the knife edge upwards between the thyroid branch of hyoid bone and larynx with a single jerk. Pull the tongue with pharynx, oesophagus and trachea backwards along the neck where previous incision was made. In small ruminants the neck and thoracic organs are removed together without cutting the trachea and oesophagus in the middle.

In ruminants while examining the lungs note the presence of exudate and parasitic worms. Check any hard areas by incision for parasitic lesions, hydatid cysts, cysticerci, abscesses, tubercular lesions or tumours. The trachea should be opened and checked for parasitic worms and bot fly larvae. The bronchial and mediastinal lymph nodes should be palpated and incised for tubercular lesions.

Examination of Abdominal Organs in Large Ruminants

Separate the duodenum between 2 ligatures at the pylorus. Detach rumen from diaphragm by hand and expose the oesophagus. Ligate oesophagus at 2 places and cut in between. Pull out the rumen with spleen and abomasum on the left side, taking care to separate the pancreas by hand which is to be taken out with liver. Divide the duodenum between 2 ligatures at the junction of its second turn with the jejunum. Free the rectum, ligate at 2 places and incise in between. Remove the whole intestines by cutting the mesenteric roots. Give a small incision in the vena cava near its entrance in diaphragm and with the help of one finger tear the vena cava through the whole length of the liver. Here observe for a thrombi or abscess in the vena cava. Remove the liver from its attachments along with pancreas and duodenum in one piece. Free and separate the kidneys and adrenals from their anterior and lateral sides. Cut the renal vessels along the medial border and carefully pull both the kidneys backwards, thus freeing the ureters from the abdominal wall. In females, separate the broad ligament of the uterus from its insertions on the abdominal wall. In males separate the already drawn penis from the ischial arch. Now the urinary and genital organs are all free. Gather all the uro-genital organs in one hand, draw them backwards simultaneously cutting the attachments. Make the incision around the anus and in the female include the vulva and safely pull out the organs, freeing their attachments. In ruminants, after cutting though the rectum, collect a dung sample. Open the rumen and note the consistency, colour and smell of the contents. If poisoning is suspected, collect samples from the rumen and abomasal content, together with 2 cm cubes of liver and kidney including the capsules. Place the specimens in clean, leak-proof jars without preservatives. Pending examination by the laboratory staff, the specimens should be refrigerated or frozen. Examine the rumen for rumen flukes (paramphistomes), if present, collect samples . Open the

Necropsy Methods 79

omasum and the reticulum and examine the reticulum for foreign bodies. Open the abomasum along the greater curvature using the enterotome. Note the condition of the abomasal epithelium and presence of any ulcers, amphistomes and bot fly larvae (in equids in stomach). Carefully check for nematodes, especially red nematodes. Examine the mucus scraped off the epithelium for very small nematodes that may be embedded in the abomasal mucosa. Put two ligatures close together between the abomasal pylorus and duodenum with a cut between two ligatures. Similarly, put two ligatures between the small intestine and the caecum at the ileo-caecal valve and cut in between. Using the enterotome, open the small intestines, wash and scrape into half a bucket of water. Examine the intestinal lining for lesions and embedded parasites. Incise and examine the mesenteric lymph nodes . Follow the same procedure to examine the caecum and large intestines. Collect and preserve the samples of all parasites seen.

In Animals with Simple Stomachs (Equids)

- Position the carcass on its right side. Severe the left foreleg by cutting all the muscular attachments that hold the leg to the chest wall and lay it flat on the ground. Similarly disarticulate the left hind leg by cutting down to the coxofemoral joint.

- Now make an incision through the skin from the anus to the chin and skin the body back almost to the vertebral column. Remove the entire upper wall of the body cavity by incising along the midline from the xiphoid cartilage to the pubis. From the pubis, continue the incision almost to the tuber coxa then forward to the origin of the last rib. Severe all the ribs near their articulations with the vertebrae using an axe or a saw. Then severe the sternal ends of the ribs from the thoracic inlet to the last rib. Carry back the incision until it joins the original incision at the xiphoid cartilage.

- Beginning at the rear, lift the severed body wall while cutting the underlying attachments, including the diaphragm. Remove all thoracic organs together by cutting the trachea and oesophagus at the thoracic inlet.

- Use a saw to cut through the pelvis, to expose the pelvic organs.

- The viscera are now exposed and can be examined. Locate and examine the epiploic foramen for herniation of the loop of intestines (important in horses).

Insert 2 fingers along the caudate lobe of the liver, a clear space allowing the two fingers in adult is the epiploic foramen.

- The stomach, trachea and soft palate of equids should be examined for bot fly larvae.

- The pelvic flexure of great colon and caecum are reflected ventrally away from the body cavity. Spleen, left kidney, left adrenals and small floating colon are removed in order.

- Duodenum is then severed behind the root of mesentery where it passes medially from the right side and is separated from the mesentery till the ileo-caecal valve is reached. (or) the duodenum can be located as follows. Follow the duodeno-colic fold mesentery from its colonic attachment near the place where the small and large colons divide, to its duodenal attachment few centimeters posteriorly.

- Remove the stomach. Examine the anterior mesenteric artery and its branches supplying the great colon and caecum, first, by removing the abdominal aorta before separating the great colon and caecum. Liver, right kidney and right adrenals are removed subsequently.

- In the horse tribe, particular attention should be paid to an examination of the blood supply to the great colon and caecum. This is accomplished by opening the abdominal aorta longitudinally and tracing and opening the anterior mesenteric artery and its branches. The presence of verminous thrombi, scarring and aneurysms should be noted.

Omnivores, Carnivores and Primate

The carcass is placed on its back. All four limbs are disarticulated and laid flat on the ground by severing the muscular attachments at the axillae and groins.

- Incise the ventral body wall from the chin to the pubis. At the sternum, the incisions should pass through the costal cartilages. Care should be taken to lift the sternum while cutting the costal cartilages so that the heart is not damaged. Now the organs of thoracic cavity are exposed.

- The ventral abdominal wall is removed completely from behind the last rib and laterally to the lateral processes of the lumbar vertebrae.

- Incise along the long axis of each mandible, hold the tongue in one hand and pull back sharply along with larynx, oesophagus and trachea freeing their attachments until the thoracic inlet is reached. The thoracic organs are removed together, with the oesophagus and trachea freeing their attachment

- The spleen and omentum are then removed and the omentum weighed if necessary. The small intestines are pushed to the left to expose the rectum, which is tied off and severed. The intestines are detached and pulled forward. The loop of the duodenum is detached from the right dorsal abdominal wall. After cutting the oesophagus posterior to the diaphragm, the liver, stomach and intestines are removed together.

- A small portion of the diaphragmatic muscle may be taken from carnivores and wild swine for examination for Trichinella spp. infection.

In the case of swine the technique is same as that of carnivores except the removal of intestinal tract. Removal of intestinal tract begins at the tip of colonic spiral which is made free along with the intestine from the mesentery.

Examination of Dead Neonates

In addition to adult necropsy some additional examinations are recommended

- Find whether breathing occurred? (do the lungs float in formalin?).

- Examine the umbilical stump - sample including surrounding tissues.

- Note the signs of dehydration / tissue moistness?

- Find out if milk remnants are present in the gastrointestinal tract?

- Check for malformations such as cleft palate, deformed limbs?

- Observe the evidence that the baby has fallen down (found on the ground, lesions, and broken limbs).

After necropsy, disinfect the necropsy site. Burry or incinerate the carcass, all remaining tissues and blood soaked dirt or waste thoroughly disinfect or incinerate all contaminated paper or plastic materials. Remove all blood and residual tissues from the instruments and tools with soap and water and then disinfect the instruments. Clean necropsy boots and apron and thoroughly wash any contaminated clothing and the external surfaces of any containers with samples.

Postmortem Examination of Wild birds

Post mortem examination is an extremely valuable tool in disease investigation and management.

- It is important to approach each carcass with an open mind, without assuming that the cause of death is known, even if there are obvious external lesions or a known on-going disease problem.

- The results of the *post mortem* examination should be used in conjunction with the history of the bird or birds and assessment of the environment.

- Obtain a detailed history of the specimen since this often gives clues as to what might be wrong.

- Perform the necropsy as soon as the bird is received because internal organs decompose rapidly and postmortem migration of parasites and bacterial invasion might occur.

- Take measurements immediately because weight, in particular, will change. Measurements are important for aging purposes and may later prove useful as indicators of specific diseases within the host population.

- Feather wear, cloacal protuberance, and brood patch will help define the breeding condition of the specimen.

- Tag the bird, and label each sample taken from it with the same necropsy number. Indicate if samples are "sterile" or "non-sterile" and the type of medium in which they are preserved.

- When the necropsy is completed, a necropsy summary can be written. Give final "diagnosis" only after the laboratory results are received.

- First prepare a smear of any exudate for Gram stain. Collect an exudate with a

Necropsy Methods

swab for bacterial culture. Then culture a portion of the lesion on mycotic medium or examine for fungal hyphae. Next, examine a wet smear of the infected tissue under a compound microscope for the presence of ectoparasites and finally the remainder of the infected tissue is divided in half and placed in 10% buffered formalin and frozen.

- Using sterile instruments, take sterile samples of any visible lesions as well as of spleen, lung, and liver. Tissue samples, swabs for culture and impression smears should be taken as tissues are examined and before they are contaminated. Tissues to be fixed should be no more than 3-5 mm thick and thinner if congested, although whole lung may be fixed in 10% buffered formalin.

- Necropsy of small wild birds is tedious. Use magnifying spectacles or a dissecting microscope. Use fine instruments (Ophthalmic tools are ideal). Iris micro-dissecting scissors, watchmaker's and micro-dissection forceps, and microprobes are invaluable. Other essential equipment includes a small pane of glass for examining the gastrointestinal tract, clean microscope slides and cover slips, sterile swabs and syringes, sterile Petri dishes and vials for collecting tissue samples, sterile plastic bags for freezing tissues, and an alcohol lamp.

- Examine the embryos and dead-in-shell chicks as well as adults.

- Perform post mortem examination in a systematic way - head-to-tail, system by system (digestive system, respiratory system etc.) or any other.

- Recognize the normal anatomy, normal appearance of organs/tissues and anatomical variation between species.

- Describe lesions/abnormalities accurately

- Keep accurate records, including a unique identifying number for each carcass and samples from that carcass.

- Avoid the use of non-standard abbreviations in permanent records.

- Keep ready, the suitable dissection instruments, sterile culture swabs, clean slides for impression smears, appropriate sterile containers and fixatives (buffered formalin and 70% alcohol).

- In the event of a large number of deaths, examine a number of individuals, representative for the range of species affected. More than one disease process may be acting at any one time and the major cause(s) of death may change during a prolonged die-off.

- Preserve samples (tissues, parasites etc.) for future reference/research.

- The person conducting post-mortem should wear a mask because humans can acquire psittacosis/ornithosis, tuberculosis, Newcastle disease, Salmonellosis and fungal diseases from birds.

- Examine the carcass and note the state of the plumage. Soiling and an unkempt appearance may indicate a long-standing illness. Check for Pox lesions on the face and legs and mite infestation on the legs.

- Examine the body for ectoparasites and the eyes and nostrils of waterfowl for flukes and leeches.

- Record the presence of flukes and lice in the crop and pouch of pelicans and in the crop of cormorants, nematodes in the gizzard of geese and lead shot in the proventriculus and gizzard of birds. Check for large nematodes which may be seen penetrating the proventriculus musculature.

Steps in Postmortem Examination

- Place the bird on its back

- If the carcass is received in a wrapper, examine inside. Note contaminants (mud, oil) and external parasites such as leeches, lice or mites which may have left the host. Collect ectoparasites (all, or a large, representative sample) and place in a bottle of 70 % ethyl alcohol. Seal the bottle and label it immediately using pencil/waterproof pen. Place one piece of card inside the container and the other on the outside of the container (NOT on the lid).

1. Record the age and sex (if known), any identifying number, bands, tags etc. Retain any rings, microchips etc

2. Weigh the carcass.

3. Measure the wing, beak and tail length and any other measurements if suggested/requested by a zoologist.

4. Before opening, examine the external surface of the carcass.

Necropsy Methods

I. Note whether the carcass is fresh or decomposed and whether it has been frozen, also whether it is intact or scavenged, and if scavenged, to what degree.

II. Permanent written records should indicate the following:

a) General appearance of carcass and plumage - e.g. wet, muddy, oiled, clean, bloody, covered in salt, b).Feathers intact, damaged or missing. If contaminated, note the extent and the areas affected.

III. Examine the carcass for a).Evidence of trauma b). Presence of parasites leeches (particularly around eyes and nares), lice, ticks etc., c).Body condition prominence of keel, whether the crop (if any, is full) and d).Check body orifices for discharges, ulceration, plaques, growths and foreign bodies

IV. Radiography is indicated to detect fractures, radiodense foreign bodies. (particularly lead shot in the gizzard, evidence of shooting, fishhooks), metabolic bone disease. Whole-body radiography may be useful for small birds to detect minor fractures, gunshot wounds, lead pellets in the gizzard.

5. Examination of Head, eyes or beak/bill, wings and legs.

Wet the feathers thoroughly with soapy water or disinfectant in water to minimize feather debris.

Head : Note any obvious injuries (trauma--bruising, lacerations, punctures, burns etc.)

Bill/beak: Note the normal appearance of the beak in the species being examined. Examine carefully for any deformities, overgrowth, scales or crusting. Cut through the sinuses, cut through the corner of the bill and examine the hard palate and the tongue. Examine the nasal cavity by incising the beak transversely at the base. With the help of a pointed scissors cut through the opening on either side to expose the infra- orbital sinus.

Inside the mouth, including on and under the tongue : Check, for any ulcers, plaques, growths, foreign bodies or burns. Expose the mouth by cutting through one corner and continue the incision through pharynx down the oesophagus to open the crop.

Ears: Raise the ear coverts and examine the external aural canal for lacerations and punctures, including bite wounds.

External nares, head, neck and body: Check for discharges from body orifices, scouring, abrasions, lacerations, fractures, dislocation.

Eyes: Note any scabs or masses around the eyes (Avian Pox). Examine the eyes, if the carcass is very fresh. Useful information may be gained by examining the eyes using indirect ophthalmoscopy. Remove the eyes carefully by dissecting with scissors and forceps. Fix in Bouin's or Zenker's solution.

Head: Cut the skin over the head and peel this back noting any bruising.
Cut through cranium on both sides from foramen magnum forwards, keeping scissors or shears perpendicular to surface to minimize damage to brain, and lift off. In small birds, curved iris scissors may be adequate to cut through the skull, while in larger birds, rongeur forceps or even an autopsy saw may be used. Examine brain *in situ*, note any haemorrhage. (N.B. An agonal bleeding may produce blood in cranial bone, but haemorrhage in meninges or brain tissue should be considered significant). Cut brain sagittally, retain samples for toxicology.

- An agonal bleeding in the skull, should be considered of pathological significance only if there are also contusions and/or haemorrhages in the overlying skin and/or intra-cranially. In older bruise the colour will be greenish. It is preferable to fix the whole brain and slice vertically after fixation.

- For access to the spinal cord, snip segments from the vertebral column and fix in formalin. Remove the cord segments once the tissues are fixed.

- In Crane: The brain should be sent for histopathological examination even in the absence of gross lesions, if there is a possibility of disease such as Eastern Equine Encephalitis or West Nile Virus Disease.

Examine the wings and the legs for any obvious abnormalities - fractures, bruises, grazed or broken skin, open wounds and swellings. Determine whether any swelling is hard (bony), soft or fluctuant.

6. Examination of body

    i. Note the condition of the skin and feathers, and the presence of any external parasites before wetting the carcass to reduce feather debris, since the state of the feathers may be relevant to the diagnosis.

ii. Check feather colour, note any fret marks, broken feathers, loss or deformity of feathers. If feather contaminants (e.g. oil) are suspected, sample feathers should be removed at this time and placed in appropriate storage containers.

iii. Examine the uro-pygeal (preen) gland (over the last vertebra), at the base of the tail.

iv. Look for brood/incubation patches.

v. Check the skin of the legs for scaling or crusting which may be due to bacterial, viral, fungal or parasitic infection. Scabs and swellings may develop with viral infection, bacterial infection or neoplastic lesions.

vi. Collect and store pieces of skin in 10% formalin.

vii. Pluck feathers along the midline as required. Place the bird on its back and open the skin from the beak to the vent in the ventral midline. Nick the skin over the breast muscles in the midline, continue the incision rostrally (forwards) to the bill and caudally (backwards) to the cloaca (vent), taking care not to damage underlying tissues.

viii. Reflect the skin from the neck, chest and abdomen. Retract the skin to expose the keel and breast muscles, ribs, and muscles over the lower coelomic cavity. Note the amount of subcutaneous fat, amount of breast muscle present, the thymus in young birds, the colour of subcutaneous tissues, any bruising, haemorrhage or other lesions, the condition of the external surface of trachea and oesophagus, any distension/impaction of oesophagus, the presence of haemorrhage, pale areas or 'rice grain' lesions in breast muscles and the presence of haemorrhage in neck muscles.

7. **Disarticulate hip joint** by giving an incision on the inner side of the thigh at the junction of the legs and the body, by cutting or breaking open the hip joint.

8. **Remove the sternum and rib cage:** Incise transversely (across the body) just below the sternum, lift the sternum upwards and, noting the condition of the air sacs, cut through the rib cage and coracoid on either side using necropsy shears or scissors (as appropriate for the size of the bird). Inspect the location and size of all organs. Examine the air sacs for transparency and note any plaques or opaque areas. The air sacs should be transparent, glistening membranes. Cheesy, mouldy masses in the air sacs and lungs may indicate infection with Aspergillus fumigatus.

Note any gross abnormalities

i.   Take care not to touch the liver, heart or lung surfaces during this procedure.

ii. Note the presence and type of any fluid or fibrin tags within body cavity, any gross abnormalities (e.g. gross haemorrhage) and the general layout and appearance of the heart, lungs, air sacs and liver.

iii. If appropriate take samples from surfaces before any contamination can occur. Note that there is no diaphragm, but there is a membranous pulmonary fold caudal to the lungs.

iv. In birds with visceral gout, a scintillating sheen of urate crystals may be visible over the viscera.

v.   Note how much fat is present and whether any fat shows serous atrophy.

vi. Repeatedly slice all parenchymal organs, to minimize the risk of missing focal lesions.

vii. Collect tissues sections of no more than 5 mm thick and place into at least 10 times the volume of 10% formalin compared to the piece of tissue.

viii. Place the air sacs on a small piece of paper before fixation (this reduces the risk of accidental discarding).

ix. Preferably fix complete sets of tissues.

x.   If gout is suspected, fix the tissues in absolute alcohol

xi. For virology, freeze tissues preferably in an ultra-low-temperature freezer.

9. **Heart:** Examine the pericardial sac, thyroid and parathyroid glands (at thoracic inlet, near common carotid arteries). There should be little fluid in the pericardial sac.

i.   Collect heart blood swabs and smears. From freshly-dead bird, blood smears should be made from both heart blood and peripheral blood to look for haematozoa.

ii. Dissect out the heart, leaving a portion of the major blood vessels intact.

iii. Weigh the heart to detect any enlargement.

Necropsy Methods 89

iv. Examine the heart for over all shape, pericardium for fluid in pericardial sac, epicardial surface for any haemorrhages, presence of worms visible under surface, thickness of ventricle walls, presence and amount of blood in ventricles, state of valves, colour of muscle-haemorrhages or pale areas within muscle or on endocardial surface.

v. Open the chambers.

vi. In small birds the whole heart may be fixed in formalin.

vii. Cut along the major vessels (e.g. aorta, jugular vein), check walls of vessels for presence of plaques / calcification.

Normal pericardial sac contains clear/translucent, trace of clear fluid. The Epicardial surface is clean and glistening. The heart is contracted and triangular. The wall of the left ventricle is two to three times as thick as right ventricle wall, with little or no blood in left ventricle and a small amount of blood in right ventricle. The valves are smooth and shiny.

10. **Lungs:** Examine the lungs *in situ*, note their colour. Normally lungs are pale pink in colour. In pulmonary edema and haemorrhage lungs appear dark red and wet.

viii. Using a gentle traction on the trachea and oesophagus, remove the lungs, while gently dissecting the lungs from the rib cage using the tip of a scalpel blade.

ix. Cut through the lungs, noting presence of any water, froth, blood, fungal infection, abscesses.

x. Using sharp scissors, cut along the whole length of the trachea (it is necessary to cut along both sides, as the tracheal rings are springy) from the mouth, and along the major bronchi, noting presence of any fluid, blood, fungal plaques, necrotic lesions, blockage and parasites (tracheal flukes, gapeworm).

There is normally a syrinx, (a structure at the bottom of the trachea formed by fusion of the lowest cartilaginous rings to form a cylindrical tympanum).

Normal lungs should appear pale pink and well aerated; the cut surfaces should be moist and glistening, often with a lot of blood present. If the bird drowned, there may be water in air sacs and lungs or just lungs if there has been a delay in examination.

11. **Liver:** Examine the liver in situ, note its size, shape, colour, any haemorrhage on surface or free in body cavity. Note whether surface is intact or split, presence of pale areas, whether lesions are flush with surface or protruding or shrunken and whether edges of lobes are sharp (normal) or rounded (enlarged). The normal liver is reddish-brown and somewhat friable; there are left and right lobes. *The liver is yellow and swollen with hepatic lipidosis, but also in neonatal chicks still absorbing the yolk sac. The liver may be diffusely yellow-orange with diffuse haemosiderosis. Examine the liver for white or yellow spots, which may indicate degeneration, necrosis or fibrosis. Multifocal pale foci (necrosis) suggests viral, bacterial or chlamydial infection. In severe anaemia the liver may be paler than normal. The presence of a whitish 'bloom' on the liver and other organs, often accompanied by a 'strawberry' appearance of the kidneys, indicates the presence of deposits of urates as in visceral gout. i).* Remove the liver, carefully cutting connections with other tissues. *If the liver is left attached to the intestines initially, until after the gastrointestinal tract has been examined, then the patency of the common bile duct can be checked by pressing the gall bladder and observing flow of bile into the duodenum.* ii).Weigh the liver (to detect any pathological enlargement). iii). Make touch preparations of the intact surface to detect elementary bodies of Chlamydia psittaci. iv).Cut through the liver several times, note colour variations, the size and shape of any lesions (pale areas, haemorrhages, abscesses etc).The whole liver should be sliced into multiple sections ("bread-loafed" to minimize the risk of missing seeing focal lesion. v).Note the size of the gall bladder. vi).Take tissue/swab for culture, impression smears from cut surface, tissue for histology and toxicology (e.g. lead levels in waterfowl) as appropriate. Plasmodium spp. and Chlamydia may be visible in impression smears from the cut surface.

12. **Spleen:** The spleen lies between the proventriculus and gizzard, on the right side. It can be found by grasping the gizzard with forceps and rotating it anticlockwise, transecting the ligamentous attachments between the gizzard and the left ventral abdominal wall as required, to visualize the spleen on the cranial portion of the gizzard. i).Note the normal size and shape of the spleen (varies between bird species). It is small and flat in waterfowl. ii).Weigh the spleen. (to detect any pathological enlargement). Enlargement may also be due to physiological causes.iii).Note the colour (normally red-brown). With viral, bacterial and some protozoal infections it may be tan and swollen. iv). Make impression smears of the cut surface, take a small piece for culture and fix the

Necropsy Methods

remainder in 10% formalin solution.

13. **Gastrointestinal tract (GIT):** Examine GIT *in situ*: Note any distension, discolouration, haemorrhage or lesions on serosal surface. Note presence and type of any fluid within body cavity. Tie off oesophagus at top of neck, cut above this, remove whole of gastro-intestinal tract to examine in the last.

14. **Kidneys, adrenals, and reproductive tract:** Examine in situ.

   i. The kidneys are paired, lobular, normally reddish-brown, found recessed on either side of the vertebral column, stretching from the lungs caudally to the last segment of the synsacrum. Note the size and colour of kidneys, presence of haemorrhage or pale areas (with renal coccidiosis, chalky white material may be present with uric acid build up).

   ii. Look for discolouration, swelling, pallor, pale or dark foci, masses, lineal white foci (indicative of gout).

   iii. The ureters are paired, running from the most cranial lobe of the kidneys to the urodeum and if normal are difficult to identify; they are visible if they are distended by urates.

   iv. Examine gonads. The female birds normally have a single ovary, the left ovary, found ventral to the cranial division of the left kidney, and a single, left, oviduct.

   v. Note the appearance of the ovary. Note the size. Note whether ovary is active (follicles present and growing) or inactive. Look for developing follicles - there should be vascular yellow follicles, varying in size in reproductively active females.

   vi. Look for lesions in the oviduct including inflammation (indicated by thickening of the distal wall of the oviduct), which indicates egg binding, presence of egg within the oviduct or the abdomen, presence of associated reaction if egg or yolk is present free within body cavity or any haemorrhage.

   vii. Note if a fully-formed egg is present.

   viii. If Salmonellosis is suspected, take ova for bacteriological examination since some Salmonella spp. can be transmitted transovarially.

ix. In Anseriformes and ratites the penile organ should be examined (in other species it is vestigial).

x. The testes in males are normally oval to elliptical, found cranio-medial to the cranial division of the kidney on either side, caudal to the abdominal air sacs and the lungs, and dorsal to the main abdominal viscera (gastrointestinal tract, liver, spleen).

15. Examine the crop and the whole of the gastro-intestinal tract, including pancreas.

i. If the contents of different segments of the gastrointestinal tract are to be submitted separately for culture, they should be tied off with string at this time.

ii. Check the serosal surfaces, note any adhesions, haemorrhage etc.

iii. Open each region (oesophagus, crop if present, proventriculus, gizzard, small intestine, large intestine, caecae, cloaca) longitudinally using scissors. Check for: haemorrhage (*N.B. agonal bleeding and post-mortem leakage must be distinguished from haemorrhage into gut*), ulceration or other lesions, presence of parasites and whether lesions are associated with these, presence and appearance of ingesta, including recognizable food items in the oesophagus, crop and gizzard, and any foreign bodies.

iv. Keep ingesta (e.g. plant pieces from the oesophagus/crop/gizzard) for analysis if appropriate.

v. Remove any parasites, preserve for identification.

vi. Examine the mucosal surface of each part of the gastrointestinal tract. The entire digestive tract should be examined for parasites, parasitic lesions and abscesses.

vii. In the gizzard, inspect the luminal surface, peel back the koilin layer and examine the glandular surface. Flush out the gizzard contents into white bowl, to examine for the presence of lead shot and other foreign bodies. Note the colour of gizzard lining and note whether it is normally adherent to the underlying tissue and whether it can be peeled out. Place a section from each segment of the GIT into 10 % formalin for preservation.

viii. Fresh samples of contents and mucosal scrapings of duodenum, ileum and caeca may be examined microscopically for parasitic ova and coccidial oocysts. Stain and examine the smears by Gram stain. The caeca may be enlarged and hardened indicating Histomonas spp. infection or coccidiosis. Do not submit sections for histopathological examination after scraping with a scalpel, as the specimen will be damaged and probably contaminated.

ix. Other structures deserving special attention are the cloaca and tendons. A generalized inflammation of the abdominal cavity and organs may indicate 'egg peritonitis'.

16. **Examine Thymus** in the jugular furrows, appearing as a series of connected lobes, grayish or pale pinkish in colour in young birds. These extend into the thoracic inlet. In some species the thymus may re-enlarge after the breeding season.

17. **Examine thyroid and parathyroid glands:** The thyroid and parathyroid glands are found closely associated with the jugular vein and the first rib, at thoracic inlet, at the level where the carotid arteries divide, with the Parathyroids just caudal to the thyroid glands.

18. **Examine the adrenals** immediately dorsal and anterior to the gonads, irregular oval in shape and normally yellowish-orange in colour. These may be enlarged with chronic stress.

19. **The bursa of Fabricius** should be examined in immature birds. The bursa is found adjacent to the cloaca/vent. Note the size and colour.

20. Before removing the thoracic viscera examine the **brachial nerve plexus**. If neurological damage is suspected collect samples in 10% formalin.

21. Examine the **sacral plexus** after removing the kidneys and the ischiatic nerve (caudal to the femur, under the medial muscles of the thigh), particularly in instances of pelvic limb paresis/paralysis (or muscle wastage indicating this) and collect tissue samples for histopathological examination. For access to the spinal cord, snip segments from the vertebral column and fix in formalin; the cord segments can be removed once the tissues are fixed.

21. Examine the **brachial plexus** between the scapula and vertebral column.

22. Examine the **sciatic nerve** between the muscles of thigh with the help of a forceps.

i. If a nerve plexus is to be submitted for histopathology, gently press flat on card and leave for a few minutes before fixing.

ii. Nerves: Take samples from the large nerves between the wing or the leg and the body wall.

23. Incise over major limb joints(knee, hock, shoulder).

   i. Examine state of synovial fluid, joint surfaces and tendon sheaths and note the colour (e.g. pale areas) or presence of any haemorrhage.

   ii. If appropriate, take swabs of synovial fluid for bacteriological examination.

   iii. Incise the leg and (later, once the carcass is being opened) breast muscles.

   iv. Note the colour (e.g. general pallor or pale areas), any haemorrhage, penetrating wounds or general mass (normal or reduced).

   v. Check whether the leg bones may be bent or easily broken. Check the feet, particularly the plantar surfaces. In web-footed birds note the state of the webs.

   vi. Skinning or partial skinning is needed to examine the muscles fully.

   vii. Collect sections of muscle and femur and store in 10% formalin.

Precautions

i. Potential hazards to human health must be considered before undertaking examination and the personnel undertaking or attending necropsies or post mortem examinations must be made aware of the potential hazards to human health.

ii. Potential hazards range from toxins on the surface of the animal (e.g. oil) to zoonotic diseases which may be transmitted through cuts, absorbed through mucous membranes or inhaled in the form of dust or aerosols.

iii. Important zoonotic diseases to consider in dealing with bird carcasses include Aspergillosis, Avian Tuberculosis, Chlamydiosis / Psittacosis, Erysipelothrix Infection, Salmonellosis and Yersiniosis.

iv. For all necropsies, protective clothing should be worn, particularly disposable gloves (which should be replaced immediately if damaged). A face mask is advisable.

Any cuts should be washed immediately with a disinfectant soap.

Records

1. All specimens should be clearly numbered and labeled. The field investigator should keep a personal record, on a checklist, of all specimens taken, how they were labeled, preserved, to whom they were submitted, and by what route. All relevant information should accompany the specimens. This information may be recorded on the Post-mortem Record and Laboratory Specimen Record forms.

2. Complete records should be kept including the species, any rings/bands/tags, microchips or other identifying information, any age/sex information, location of origin, whether captive or free living, who presented the carcass for examination etc.

3. The general state of the carcass should be reported - fresh or degree of decomposition, whether whole or not, whether refrigerated or frozen before examinations, time from death/the carcass being found.

4. A full record of the necropsy findings should be kept, preferably set out on a standardized form. Necropsy records need to be properly organized and retrievable.

5. Necropsy records should be properly linked to other records regarding the individual, such as its health records and history (for birds within a collection; or for free-living birds whether the carcass was found alone or part of a die-off etc.).

6. A record should be kept of any samples taken, for what purpose (histopathology, bacteriology, virology, parasitology, detection of pesticide or heavy metal residues etc.

7. AVOID using non-standard abbreviations. Standard abbreviations (e.g. "NE" for "not examined", "NLD" for "no lesion detected") should be defined on the necropsy form.

8. All samples should be properly labeled with identifying information which will match the samples with the necropsy form and other appropriate records.

9. Records and good identification of specimens are particularly important when investigating deaths which may be malicious or may lead to prosecution or compensation claims (e.g. suspected poisonings, oiling etc).

10. If legal proceedings are possible, preferably have a witness present throughout the necropsy to confirm the accurate recording of findings and to witness labeling and sealing of possible forensic material. Additionally, keep the carcass deep frozen as well as suitably preserving the organs.

11. Photographs provide a useful record of the appearance of organs and lesions and can be an important aid to diagnosis.

4 | Autopsy Report

It is a statement of facts relating to (1). the observations made at autopsy and (2). after microscopic examination of tissues.

These facts together with the results of special laboratory examinations (microbiological, parasitological and chemical) lead to conclusions about the nature of observations i.e. autopsy findings and to the ultimate diagnosis.

The primary rule for making pathological observations and descriptions is " that the observed facts must be strictly separated from interpretation of these facts".

The gross description begins with the description of the animal. These are Species, breed or type, sex, age, weight and identification marks. If age is not known it should be estimated and facts mentioned. The description also includes a statement of the nutritional condition. viz. Obese, good, moderately good, poor or inanition. The terms obesity and (the other extreme condition) inanition are objective terms. Obesity is an abnormally great accumulation of fat in the body stores. Inanition is a state of emaciation characterized by serous atrophy of adipose tissue, especially evident in the coronary grooves of the heart and the bone marrow and some degree of atrophy of other body tissues, particularly obvious in the skeletal musculature. The other evaluations of the nutritional state are arbitrary. With experience the observer will be able to establish a standard scale. The details regarding the stage of oestrus, gestation, lactating or dry also form a part of the general description of the animal and are recorded in the appropriate place in the report. Description regarding the degree of rigor mortis and cadaveric changes are also mentioned. These will have a major value when interpreting the results of histopathological and special laboratory examinations. In addition one should also look in for other changes involving the body as a whole viz. dehydration, discolouration, anaemia etc. Observe also the natural orifices and wounds on the body.

When the animal is being examined, the pathologist should observe whether there are deviations from normal, if so the nature of such deviations. Generally one should observe the deviation / lesion involving whole organ first and then its parts. The same pattern will be followed in the pathological description. The report is confined to a description of deviation from the normal. But at times it is necessary to state definitely that an organ or structure has a normal appearance. For example, if there is an obvious discrepancy between the observed and the expected pattern of pathological changes or if one wants to emphasize the fact that the changes are focal in an organ or limited to one member of a bilateral structure.

The gross description, whether of an entire organ or structure or of focal pathological changes, is statement of size, shape, colour, consistency, appearance on cut surface and relation to surroundings. The description of focal lesions also requires a statement of (1) anatomical relation and when numerous, (2) the number (if countable) and (3) the pattern of distribution (evenly distributed, in clusters etc).

Size: It is an objective quality and best indicated by actual measurements like weight, linear dimensions or sometimes volume whichever is most appropriate to a particular situation or lesion. The words such as enlarged, small, shrunken, dilated and contracted do not have absolute value. These represent interpretations and are used only in a comparative sense or when more accurate reference to size are inapplicable or inadequate to convey the over all impression.

Focal changes in an organ when numerous, often vary in size and shape. In such cases one should state the range and model characteristics.

Shape: The shape of irregular, three-dimensional structures and lesions can rarely be described accurately. One can be content with stating the overall appearance viz. pyramidal, oval, egg shaped, stellate, spherical, ovoid, conical, triangular, flattened, nodular, tortuous, spindle, wedge shaped, mushroom shaped, dome shaped, whorled etc.

Referring the size and shape or structure of lesions to the dimension and form of common objects may sometimes aid in communication but should be practiced with discretion, if it is to be meaningful.

Colour: It is not difficult to describe a colour provided one refers to the names of the colours in the spectrum i.e. red, orange, yellow, green, blue and violet together with black, white and gray. Combination of any 2 adjectives will cover all the colour tints likely to the encountered (e.g. green blue means more blue than green). Light (pale) and dark (deep) will suffice as modified meaningful adjectives.

The surface may be hairy, smooth, covered with exudate, irregular, eroded, rough, elevated, depressed, glistening, dull, scaly, membranous, pitted, undulant etc.

Consistency: is referred to as soft, hard, firm and resilient (elastic) with the usual modification of slightly, moderately or very.

Description of fluids / exudates: Watery (serous), viscid (mucoid) with appropriate modifying adjectives can be used. Exudates may be pasty, crumbly or inspissated. A statement of colour is modified by "clear" or "turbid". The material in suspension is described as such.

Amount: This is expressed in volume if possible.

Appearance of cut section of organs: This includes statement of appearance and distribution of changes, their relation to anatomical landmarks, colour and consistency. It is useful to distinguish between "generalized" and "diffuse" i.e. between distribution throughout the whole organ from distribution within one part of the organ Ex. Between generalized pulmonary emphysema vs diffuse emphysema of apical lobe. The characteristics of the cut surface of focal changes- whether homogeneous, special structural details, capsule etc and again the colour and consistency- are described as they appear for most of the focal changes, if they are numerous and details are given for only the important variations from the general pattern. A description of relation to surrounding tissues and structures is given for focal changes and includes such details as sharp or diffuse transition, compression or infiltration, haemorrhagic or hyperaemic bordering zones and so on.

In case of tubular structures, one must note whether it is patent / obstructed / dilated / obliterated / narrowed /branched /or a diverticula present.

Microscopic observations: The principles of description of microscopic findings are the same as those for gross autopsy description. Facts and not interpretations are given. The descriptions progress from the general to the specific changes noticed. The results of special staining or of histochemical reactions may be given in the last or may be woven into general description if the basic findings can be described more precisely and concisely. In case of tumours, the general architectural pattern is described first, and then the details of cytology and local variations are given and finally the relation to the surrounding tissue if present are detailed. The various terms, pyknosis, necrosis, hydropic degeneration, acanthosis are interpretations. These can be used for conciseness and clarity. The quantitative evaluations in microscopic descriptions are expressed by such terms as limited or extensive, slight, moderate or severe and in the case of numerical quantities to few, sparse, moderate or abundant. The terms severe and extensive are not synonyms.

A good pathological description should be neat and efficient. A precise, concise and lucid description gives confidence in the reliability of the observations and the conclusions drawn. While describing, avoid obscurity, too long wordy and superfluous description. It is always good to use simple, grammatically correct sentences, containing exact vocabulary arranged in a logical sequence with facts.

Part II

Wildlife Forensics

1 | Introduction

Wildlife means the native wild fauna and flora of a region and includes any animal, aquatic or land vegetation which forms part of any habitat. Wildlife crime can be defined as illegal taking, possession, trade or movement, processing, consumption of wild animals and plants or their derivatives in contravention of any international, regional, or national legislation(s). Infliction of cruelty to and the persecution of wild animals, both free-living and captive are also at times added to this definition. *Wild animals and plants are the victims of any wildlife crime.* It is a growing international problem that threatens the survival of many species.

Wildlife forensics is a relatively new field of criminal investigation which uses a multi disciplinary scientific team that includes a variety of physical and biological sciences. Analytical chemistry, firearms and bullet analysis, photo documentation, computer analysis, trace evidence analysis, along with DNA, morphological analysis and pathological evaluation all work together. It uses scientific procedures to examine, identify, and compare evidence from crime scenes, and link this evidence with a suspect and a victim, which is specifically an animal. Killing wild animals that are protected from hunting by laws, (called poaching), is one of the most serious crimes investigated by wildlife forensic scientists. Other crimes against wildlife include buying and selling protected animals and buying and selling products made from protected animals. Knowledge of and respect for the collection needs of other disciplines will go a long way in putting together an effective analysis of available evidence.

Very few guidelines are available for wildlife pathologists who perform necropsies on such cases. When evidence regarding a necropsy is required in a court, there is usually examination of both the pathologist's formal training (academic credentials) and experience in forensic examination to establish his or her ability to serve as an expert witness. Whenever possible, forensic necropsies should be performed by a trained pathologist who has experience with the species being examined. Crimes against wildlife encompass a wide range of offences, from the illegal trade in endangered plant and animal species to the persecution of birds of prey or the cruelty inflicted on some wild animals for sport. Combating wildlife crime is essential to preserve our natural heritage. Products that are made from animals are also of interest, including leather goods and medicines, especially those from Asia.

2 | Crimes against Wildlife

Wildlife crime occurs at a global scale. Criminal activities involving illegal international wildlife trade can have far wider implications on disease transmission, invasive species introduction and natural resource exploitation. The illegal trade in plants and wildlife can adversely affect people's livelihoods and the resources they depend upon, as well as threaten already dwindling populations of endangered species. Removal of species from the wild can also upset the delicate balance of nature in the ecosystems where they live.

Wildlife crimes differ from other crimes like murder, theft, drugs etc., in following aspects. (i).Wildlife crimes are location specific. To commit a wildlife crime, the offender has to invariably go to the place where the targeted wildlife is available.(ii).Earlier these acts were not criminal offences until enactment of the Wildlife (Protection)Act, 1972 (which are now construed as a wildlife offences). Hunting was once an act of valour and a royal pass time. (iii).Public at large is not affected or disturbed by the wildlife crimes unlike the case of conventional crimes like murder, theft or robbery, every incident of which infuses a sense of fear into the minds of people and hence try to prevent such crimes. This does not happen in case of wildlife crimes.

Reasons for Illegal Trade in Wildlife

- Huge profits earned by the traders.

- Highly lucrative trade (due to Low .risk and low penalties).

- No stigma is attached to the offenders who commit wildlife crimes unlike other conventional crimes.

- Readily available wealthy markets for illegal trade of wildlife in Asia, Europe, USA and the Middle East.

- Craze for ornaments made of animal body parts (ivory, tiger teeth/bones).

- Use of animal body parts or plants in traditional medicines.

- Keeping the skins or horns or antlers as status symbols, and cultural beliefs or even, superstitious beliefs.

- It is a growing international problem that threatens the survival of many species.

Wildlife crime takes many forms some of which involve extreme cruelty carried out by organized gangs. These are

- Illegal trade in endangered species.

- Killing, injuring, taking, disturbing etc. wild birds and wild bats.

- Trapping of live birds such as finches for export.

- Taking/possessing/destroying wild birds eggs/nest disturbance.

- Badger persecution / badger digging.

- Illegal trapping/snaring of wild animals.

- Illegal hunting of wild mammals.

- BUS H -M E A T : Bush meat, sometimes called 'wild meat' (in the Caribbean, for example) can be defined as 'wild animals killed for food' and the species involved can range from grasshoppers to guinea fowl and gorillas.

- Illegal trapping, shooting, snaring or poisoning of birds or animals.

- Damaging protected sites.

Crimes Against Wildlife

- Illegal poisoning of wildlife.

- Disturbing cetaceans.

- Stealing wild plants digging up, or in some cases picking, wild plant(Wild plant theft).

- Illegal hunting and poaching-- poaching of deer, game or fish.

- Destruction/ damage to habitat.

- Taking of freshwater pearl mussels from rivers.

- Introduction to the wild of non-indigenous species.

- Poaching of protected or endangered species. It is done locally to provide the food for the poacher and his family. Large scale poaching leads to illegal trade in wild animals and their derivatives and has serious effects such as decimation, even disappearance of target species from the area. Poaching occurs in much of the world. It is extremely difficult to investigate and prosecute due to the nature of evidence available.

- Taking (including killing) alive or dead, of protected wild animals or plants without appropriate authority.

- Unlawful persecution and killing of protected species because of human animal conflict over land and resources.

- Killing or causing injury to protected species by physical means (e.g. draining marshland) or by polluting habitat (e.g. discharging oil into the sea).

- Causing disturbances to specially protected species, their nests or habitats.

- Possession of protected wildlife (animals or plants) or their derivatives without appropriate authority.

- Importation, exportation or sale of protected wildlife without appropriate authority.

- Offences relating to the maintenance of protected wild animals in captivity.

In general it is an offence

- to kill or injure any protected wild bird / animal.

- to capture or keep (alive or dead) any wild bird / animal without appropriate authority.

- to destroy, damage or take the nest of any wild bird while it is in use or being built.

- to destroy or take the egg of any wild bird to disturb any wild bird in, on or near a nest containing eggs or young - to disturb the dependent young of such a bird.

- to sell or advertise for sale any wild bird or its eggs / animal or anything derived from it.

- to destroy, damage or obstruct the animal access to its place of shelter and

- to disturb the animal while using its place of shelter.

Many species, including Otters, Badgers, Seals, Red Squirrels and some Butterflies are given legal protection. The more general law controlling when, where and how one can hunt, shoot or fish is also relevant.

Humans may injure or kill wildlife in a variety of ways sometimes intentionally, sometimes by neglect and sometimes by accident. The task of the investigator is to find out with the assistance of a veterinary pathologist, whether or not the injuries or death were likely to have been human induced , caused by other species or the result of such a factor as hyperthermia, hypothermia or starvation. Assault on wild animals can be malicious or accidental and can be further divided into (1). **Physical** : trauma, excess heat or cold, immersion in water or other insults.(2).**Sexual:** attempted copulation and damage to reproductive organs and (3). **Psychological:** teasing, taunting or threatening, deprivation of company and unsuitable social grouping.

One important aspect in crimes related to wildlife is the trade in the wild animals, plants and their parts, products and their derivatives. India is an importer, exporter and a conduit for wildlife. In recent years the trade in the medicinal plants and wild plant materials has grown tremendously along with the growing fear of piracy of genetic resources and hijacking the traditional knowledge. In wildlife crime the evidence tends to be part of your victim (related parts or products to the species)

Crimes Against Wildlife

Most Commonly Trafficked Fauna

- Elephants – ivory for jewelry
- Tropical fish for food
- Bears – gallbladder for medicine
- White Rhino - horns for medicine, skin
- Tibetan Antelopes – wool
- Sea Turtles - meat and eggs for food.
- Sharks – fin for cooking soup
- Wild birds – exotic pets
- Tigers – skin for trophy, Claws, bones, Whiskers and every part of the body
- Wild birds and their eggs, Wild animals as pets and for meat.
- Leopards killed for Claws, bones, skin, Whiskers and other parts of the body.

Asia and the Pacific

Asia is a hotspot for wildlife trafficking. This region is home to tigers, Asiatic lions, and snow leopards and is a target of international crime networks. The trade is a multi-billion dollar a year industry. Weak enforcement, government corruption, and weak border controls make the trade a strong and profitable force in the region.

Africa

Important animal species in Africa, such as the African elephant and the gorilla, are increasingly threatened by animal traffickers. Wildlife trade in Africa is stronger than ever, mainly because regulations are impossible to enforce. Most African nations have unstable governments and weak economies.

Latin America and the Caribbean

Latin American countries are very significant in the wildlife trade. Millions of species were removed from the Amazon Rainforest in Colombia, Brazil, Peru, Venezuela,

Ecuador, Bolivia etc and are smuggled to North America and Europe. Wildlife crime in this region is large because many drug lords are also involved in species trafficking. Police forces are not strong enough to combat the problem alone, and laws to fight this kind of crime are not tough and are rarely enforced.

There are 3 main ways in which wild animals, free living or captive are involved in litigation.

1. Be the cause of an incident - Injure a person. Damage property or spread an infectious disease.

2. Be the victim - If killed, injured or poached, exported illegally or treated inhumanely.

3. Animals as sentinels - Provide information that is relevant to an incident since they are present when a crime was committed and their hair or other material remains as evidence.

Animals of different species singly or in groups can cause injuries, death and financial loss. Wildlife forensics can involve domesticated animals, as well as humans (A pet dog may be responsible for killing or maiming a protected species).

As sentinels

Wild animals may form part of forensic investigation Ex. Birds or rodents may be found dead in a farm building where fire or explosion might have occurred. Clinical examination, necropsy and laboratory test may indicate when and as to why the accident has occurred and evidence as to what might be responsible for the accident—traces of explosives or chemicals. Animals also provide valuable information about the hazards of certain poisons such as catastrophic poisoning of birds of prey by chlorinated hydrocarbons and the precipitous decline over the past 2 decades of Asian vultures.

Human induced injuries in wildlife are noticed in

- War and civil disturbance

- Disasters--human induced and natural disasters

Crimes Against Wildlife

- Bush meat (killing wild animals for food)

- Killing animals for traditional medicines

- Hunting, shooting and fishing. Legal hunting and illegal poaching, both contribute for bush-meat trade.

- Wildlife sport

- Releases of oil and gas on land and into the sea

- Accidents involving wildlife, captive wildlife zoological collections, aquaria invertebrates (key part of biodiversity)

- Animals in entertainment, circuses, and

- Wild animals in research

Wildlife crime is a response to poverty and a need to obtain food, not driven by greed or **criminality**.

If the animal is the victim, the legal cases fall into the following categories

i.   Animals found dead under unusual unexpected or suspicious circumstances. In this a dead animal or a group of animals is investigated with a view to determine the cause, mechanism and manner of death.

ii. Animal is alive but exhibits unusual, unexpected or suspicious clinical signs. Investigation relates to live animals that are found in a sick, injured or incapacitated condition and the circumstances appear unusual or suspicious.

iii. Animal's welfare is or has been compromised. There is a need to determine whether an animal is or has been subjected to unnecessary pain, discomfort or distress. Cruelty may be alleged. In some, the ill-treatment of a wild animal may be linked to abuse of livestock or domestic pets or violence to humans

iv. Animal appears to have been taken, killed or kept in captivity unlawfully.

v.   These cases are termed as wildlife crime i.e. activity directed at wild animals, plants or their habitats which constitute an offence, under national, regional or international law.

Techniques in Wildlife Crime Investigation

1. **Clinical work**: Examination, diagnosis, sampling and treatment of animals. Diagnosis and treatment require by law the services of a registered veterinary surgeon. It includes radiology and other imaging techniques and laboratory investigations.

2. **Identification and morphological studies:** Involves gross, microscopic and DNA based techniques for species determination using bones, skin, hairs, feathers, scales and derivatives.

3. **Pathology:** PME of whole carcasses, organs and tissues from dead animals and samples from live animals. Can include various supporting tests like radiology and other scanning techniques and laboratory investigations.

4. **DNA Technology:** A very important component of many investigations. Employed for species, sex and parentage determination, geographical origin linking an animal to human and other animal or object and other investigations.

5. **Stable isotope techniques** to find out geographical origin.

6. **Toxicology:** To find out the natural, malicious or accidental poisoning.

7. **Crime investigation:** A standard investigations involving police and specialist. Involves studies on ballistics, chemical assay, soil analysis, finger prints, tyre marks etc.

8. **Crime scene studies:** May require the experience of field work and knowledge of natural history.

Methods in Wildlife Forensics

Most cases involve a combination of the following.

1. Visit to the scene of alleged crime and an assessment of what is found or seen.

2. Information gathering primarily by interviewing people who are or are believed to be involved in the incident or may have relevant information.

3. Examination of live animals

Crimes Against Wildlife

4. Examination of dead animals

5. Collection and identification of evidence including derivatives and samples for laboratory testing which may necessitate a certain amount of field work.

6. Correct transportation of specimens to the laboratory for testing

7. Laboratory tests sometimes in the field

8. Production of reports

9. Appearance in the court

10. Correct storage and presentation of evidence

11. Retention of material for further court proceedings or for reference purposes.

Some Questions that may be asked about the Animal

- Species/ subspecies/hybrid
- Sex/entire/neutered
- Sexually mature/fertile/in oestrus
- Pregnant/gravid
- Lactating/incubating
- Geographical origin
- Migrant/resident
- Captive or free living for how long
- Health status
- Age of skin wound, bruise, fracture or other wound (for carcasses or tissue-how long dead)
- Origin of animals and their derivatives
- Live animal
- Dead animal
- Hairs, feathers, scales including shed skins (slough) of reptiles
- Horns, antlers and hooves
- Blood and tissues
- Eggs, embryos, foetuses and placentae
- Ingesta, faeces, urine and saliva
- Bones and teeth including elephant ivory
- Macro and micro parasites.

Crime investigation is an important part of wildlife forensics and involves, deserts, mountains, forests and marine environments. Wildlife forensics presents many challenges and those involved in investigation may face the threats and physical dangers especially when working in the field.

Many wildlife crime investigations will rely on evidence from professional or expert witnesses. The expert witness's primary responsibility is to the court, even if they are called and paid for by one of the parties to the case, and they must remain independent of their party's vested interest. It is essential that the person or agency undertaking the work is suitably qualified or experienced. An expert witness is expected to have a sound, current and practical knowledge of the subject matter, based on actual clinical or practical experience. Furthermore, they should be able to provide accurate and robust evidence that can withstand challenges that may be made in court. Increasing attention is being given by the court to the quality of work undertaken by forensic practitioners. The expert witness is likely to be subjected to rigorous sometimes hostile cross examination. If the expert witness has not presented the case properly, the meticulous forensic examination may be wasted.

A 'professional witness' is one who by reason of some direct professional involvement in the facts of a case is able to give an account of those facts to the court. Thus a professional witness is a witness of fact, who is also professionally qualified, such as a veterinary surgeon. A veterinary surgeon, as a professional witness, may be able to undertake a perfectly competent postmortem examination. However, the court may be best served by using a veterinary pathologist, perhaps with experience in a particular animal group, who is able to provide specific expert evidence. Consequently the choice of specialist or expert may need to be considered at a very early stage.

Qualifications as an Expert witness

Training(formal schooling).
Experience-related to area of expertise (how many similar cases have been examined-forensic necropsies done by species).
Special training and experience (residencies, seminars, personal tutorage).
Publications and presentations.
Professional recognition and activities and Court room experience.

Witness of fact - May only answer questions related to what was personally seen or experienced (can only testify in a court of law as to what he saw or heard)

Expert witness — May offer an opinion or interpretation based on training, knowledge or experience (may give an opinion as to the meaning or otherwise, provide testimony within his area of expertise).

The judge will determine if you are qualified as an expert witness based on the evidence of your qualification. The jury will determine if you are a credible witness based on your testimony.

Crime investigation necessitates the correct collection, processing and storage of evidence for possible production in the court. Standard techniques , following established protocols, meticulous investigation, proper selection, labeling and transfer of samples, reliable chain of custody, accurate record keeping and quality management at all stages of forensic investigation are essential. An essential starting point in any forensic investigation is the proper collection of evidence and avoidance of potential cross contamination. An assessment should be made to establish what would be the most useful type of evidence to obtain in relation to the offence under investigation.

Animal Protection Laws (India)

Central laws
The Prevention of Cruelty to Animals Act,1960
The Performing Animals Rules,1973 and The Performing Animals (Registration) Rules, 2001
The Prevention Of Cruelty To Draught And Pack Animals Rules,1965
The Transport of Animals Rules,1978
Transport of Monkeys
Transport of Cattle
Transport of Equines
Transport of Sheep and Goats
Transport of Poultry
Transport of Pigs

The Prevention of Cruelty to Animals(Slaughter House) Rules,2001
The Experiments on Animals (Control And Supervision) Rules,1968

3 | Forensic Analysis of Wildlife Crimes

The field investigator has certain responsibilities when investigating animal "crime scenes". The veterinarian may or may not be involved in the scene investigation. The "scene" may be a natural mortality or a true violation of wildlife conservation laws or other animal related laws. In such a case, the "crime scene" must be protected until fully investigated to, properly preserve and collect the physical evidence. The evidence then must be submitted for evaluation to qualified investigators using approved evidence chain of custody. The field investigator must then reconstruct and document the events of the crime scene and pass the information on to the pathologist or other investigators.

Linking a suspect to a crime scene and a victim is an essential part of any criminal investigation. Wildlife officers have a wide variety of forensic techniques available for use in the investigation and prosecution of wildlife crime. An essential prerequisite to any wildlife legal case is the correct identification of an animal's species. In addition to being asked to identify species and individual animals, forensic scientists must often answer questions about sex, parentage, the number of individuals in a sample, and the time of death (although that can't be determined on trophy mounts and frozen carcasses).In some countries , it is based on the traditional techniques such as gross, morphological examination of body parts rather than more sophisticated and expensive DNA based molecular methods. Forensic entomology is an important

Forensic Analysis of Wildlife Crimes

aspect of wildlife forensic work and can provide valuable information about the circumstances of death and the movements of both the perpetrator and the victim.

The wildlife forensic scientists may have to distinguish if a piece of leather on a watchband is made from a protected animal, like an elephant or a zebra, or if it comes from a non-protected animal, like a cow or a horse. They must be able to determine if a medicinal powder contains the pulverized remains of a protected animal, like a rhinoceros, a tiger, or coral. They must be able to differentiate between the roe of protected fish from farm-raised caviar. A variety of scientific techniques allow wildlife forensic scientists to answer these types of questions. Techniques similar to those used in a police crime laboratory are used to identify and analyze parts of animals as well as bullets, shot casings, paint chips, soil, and fibers found at the crime scene. Experts in fingerprinting, ballistics , soil analysis, and hair comparisons examine evidence visually and with microscopic techniques.

The services of the following personnel are utilized in the wildlife crime investigation to identify the species of animal or plant from the evidence collected, whether it is the entire organism, parts or pieces or even products such as clothing, jewellery or processed meats and determine the cause of death.

1. The Veterinary Pathologists examine carcasses for wounds in order to determine how the animal died and to distinguish natural death from human killing.

2. Experts in the morphology (Morphologist), or the form of animals can identify the species, and sometimes subspecies of animals found at crime scenes. They can often determine the age and sex of animals as well as the time-since-death by careful observations of feathers, skulls and skeletons.

3. Chemists (Analytical chemists) may be asked to identify poisons and pesticides, characterize the contents of Asian medicines, and provide species identification, when possible.

4. Molecular biologists use protein and DNA analyses to provide information about the identity of a sample. Genetics can be particularly useful when the sample is very small or unidentifiable from its morphology. Some answers that genetic tests may provide include identification of species, characterization of the familial relationships between animals, and evaluation of two different samples in order to determine if they originated from the same individual. In addition,

geneticists may be able to provide environmental information about an animal.

5. The services of Criminologist are utilized for the identification and comparison of wildlife-related evidence. The different units of criminalistics unit evaluate the physical evidence collected from the crime scene such as ballistics evidence, tool markings and soil, questioned documents (falsified permits, licenses and other documents) and animal and human prints. Both biology and chemistry units are integrated into trace evidence assessments. In addition to identifying the species, and individual animals, sex, parentage, the number of individuals in a sample and the time of death are also determined.

Different types of evidence, both biological and non-biological, are used in the investigations of wildlife crimes. The investigation of any wildlife offence seeks to answer the questions of who, what, where, when, why and how.

The issues which feature prominently in wildlife cases are
i). the identity and the provenance of the specimen in question , ii).cause of death or injury as well as iii). connection of the suspect to the crime scene.

Types of evidence

Biological evidence	Non-biological evidence
1. Species' identification	1. Footprint and tyre tread impressions
2. Individual identification	2. Ballistics
3. Population identification	3. Fingerprints
4. Cause, manner and time of death	4. Examining documents of questionable origin

Examples of evidence items examined are

- Any part of the animal including blood samples (ideally fresh or dried condition).

- Tissue samples (frozen)

- Whole carcasses.

- Bones, Teeth ,Claws, Talons, Tusks, Hair, Hides Furs and Feather.

- Stomach contents.

- Materials used to kill or harm animals, such as Poisons, Pesticides.

- Projectiles (e.g. bullets, arrows), Weapons (e.g. rifles, bows, traps) and

- Products that are made from animals are also of interest, including leather goods (e.g. purses, shoes, and boots) and Asian medicinals (e.g. rhino horn pills and tiger bone juice).

1. Role of Pathologist

Forensic necropsies should be performed by pathologists with formal training and experience, because these credentials will be examined if a case reaches the court. When evidence regarding a necropsy is required in the court, there is usually examination of both the pathologist's formal training (academic credentials) and experience in forensic examination to establish his or her ability to serve as an expert witness. Hence whenever possible forensic necropsies should be performed by a trained pathologist who has experience with the species being examined. The veterinary pathologist (uniquely qualified by training and experience) will assist state and federal wildlife enforcement officers in solving crimes involving animals by providing a cause of death or other pertinent facts. He will be recognized as an "expert witness".

The pathologist works on intact carcasses.
The steps in a forensic pathological evaluation include:

Field investigation (i.e. Crime scene investigation or "history").
Proper collection, documentation and shipment of evidence.
Examination of evidence.
Documentation of observations.
Collection and documentation of additional "trace" evidence.
Further processing and evaluation of "trace" evidence.
Submission of official report with conclusions.
Attorney or court presentation.

By thorough and careful observation , the pathologist will recreate, interpret and document the terminal events in a carcass of an animal submitted for forensic examination. If it was poisoned, then chemist will try to determine what type of poison. A tissue sample is sent to geneticist to determine if it is a wolf or a dog, and often to Morphologist as well to determine if it has lived in captivity. All sections of the laboratory, work together as a team to try to assist law enforcement in protecting wildlife. Pathologist can also determine the difference between an arrow wound and a gunshot would. Pathologist uses the wound characteristics to determine the type of wound and X-rays to determine the true cause of death.

Types of cases Wildlife pathologists deal with

1. Assessment of welfare (Animal abuse and neglect)
2. Poisoning / sudden deaths
3. Zoonoses
4. Human injury,
5. Damage / Insurance
6. Fraud
7. Speciation
8. Liability
9. Nuisance
10. Wildlife crimes

The responsibilities of a pathologist in forensic cases are

1. To process evidence in a manner consistent with forensic principles.

2. To provide objective scientifically based complete pathological valuations.

3. To document and preserve evidence integrity.

4. To maintain confidence regarding the case and

5. Be available to provide professional quality expert testimony. It is the responsibility of a pathologist to reconstruct the event as best as he can within the constraints of the available evidence without assumption. The pathologist must be true and unbiased since he is the finder of facts.

Duties of a Pathologist in forensic investigations are

1. To identify the animal as to species, sex and age and collect physical evidence such as bullet fragments or toxic residues.

2. To determine what the law enforcement wants to know and whether it can be delivered.

Forensic Analysis of Wildlife Crimes

3. To determine the cause and nature of death of the animal being examined and determine the time sequence in which the events occurred.

4. To conduct examination of animals and give unbiased opinion (to avoid improper influence on interpretation of findings).

5. To maintain admissibility of evidence with documentary evidence and findings.

6. To protect personal professional credibility.

7. To anticipate what might legal system, enforcement, defendant or press wants to know.

8. To assesses the general health of the animal and the presence of preexisting conditions that may influence its death and collect information and perform ancillary tests to rule out alternate explanations (to show that the animal was in good health and not suffering from an infectious disease at the time of death).

9. To provide law enforcement and legal system with objective data to make decisions (prosecute and render verdicts).

The pathologist performs the following in all forensic Necropsy cases

1. Conducts detailed examination of animals, describing the detailed gross, histological and other findings.

2. Retains the samples for toxicology and microbiology even when not required.

3. Documents the conversation made with other experts.

4. When severe trauma/ bullets are involved, takes radiographs.

5. Photographs the lesions and conditions of carcass.

6. After necropsy freeze back carcasses / body parts-Preservation.

7. Consults the experts in other fields to perform analyses (Conclusions about the ballistics or projectiles are best drawn by forensic firearm specialist).

8. Provides the list of documents indicating the receipt of specimens--sign and date log, Photos to support identity of specimens, evidence tags (accession number for the laboratory, signature of the pathologist, date and scale), marking the photos that have proven link to the case(people coming into possession of the specimens are links in a chain).

9. Documents chain of custody.

10. Maintains and keeps the specimens in secure areas to avoid contamination, mix ups and tampering.

11. Submits chain of custody log.

12. Maintains secure control over specimens and collected information till they are handed over to law enforcement officials or disposed off.

13. Written records—Describes the findings concisely that is understood by non-medical persons. Keeps relevant documents (photographs, tags, frozen / fixed tissues/samples) to support conclusions.. Relies on the information of veterinarian, owner and the police regarding the significant features in animals environment.

14. Collects samples (DNA or bone samples)and sends them to identify the species (dog vs. wolf), genetic relationship (genetic laboratory), presumed human semen (state crime laboratory); bullets (state ballistic expert), unusual poisons (toxicology laboratory) and the performance enhancement drugs (drug testing laboratories).

The pathologist usually faces the following challenges while examining the dead animals.

1. Postmortem changes in the dead animal, especially in the wild animals due to impact of temperature, body condition and predation as well as due to decomposition of the carcass.
2. Autolysis and bacterial putrefaction.
3. Skeletalisation.
4. Adipocere
5. Mummification.

Questions answered by pathologist are

1. Could spontaneous disease account for sudden death of the animal?

2. Was the animal shot? If shot how, with what. He also provides bullet to state crime laboratory.

3. Was the animal starved.

4. Did vaccination/surgery/medication make animal ill.

Forensic Analysis of Wildlife Crimes

123

5. Type of poison involved? If so what. (sends tissues for chemical laboratory for report).

6. Was surgery responsible for death of the animal.

7. Will lesions noticed agree with owner's / hunter's story.

8. Is the animal progeny of specific sire.

A major difference between an ordinary necropsy (done in a wildlife disease diagnostic laboratory) and a forensic necropsy relates to the completeness of records that are kept about the procedures and the results obtained. The pathologist may appear in the court to testify several years after the necropsy was done (when it reaches the court) at which time he may have a faint memory of the case. He must necessarily depend on the written report available. The final necropsy report will form the basis of any examination that occurs in the court. The description of necropsy report should be complete, clear and phrased in non-technical language. Negative and positive findings should also be recorded. Abnormalities or lesions should be described in absolute units (centimeters, grams etc) rather than in relative terms (enlarged, smaller than normal etc). Outline drawings of location of injuries/ abnormalities may be prepared. Location of injuries/abnormalities should be measured and recorded in relation to the readily identifiable anatomic landmark. The original notes of necropsy findings should be retained by the pathologist until the possible court proceedings are completed. The final report should be read carefully by the pathologist and each page signed before it is sent. The report is confidential and should be given to the authority who requested the necropsy examination. All documents including necropsy report, photographs and results of ancillary tests and physical specimens must be retained in the pathologist's custody in a secure manner until the case goes to the court or until the enforcement officer advises that the case has been closed and the information is no longer required.

Veterinary Pathologists are responsible for determining cause of death of an animal carcass submitted as evidence. This is accomplished through necropsy (autopsy) protocols involving a search for lethal wounds caused by bullets, arrows, spears, and traps (in order to determine how the animal died and to distinguish natural death from human killing). The animals dying of natural causes or even at the paws or teeth of a predator need to be ruled out. Partial or complete skinning of the carcass may be required to document trauma patterns, bullet or arrow wounds, predator and scavenger damage, etc. A comprehensive toxicological investigation of blood, urine, tissue, and stomach/crop contents to eliminate or confirm a poison or a contaminant as a cause of death should be undertaken and a professional evaluation of the underlying health of the animal prior to death may also be carried out. Examining a carcass for evidence of electrocution is also the job of the pathologists. In the process of conducting these examinations, the veterinary pathologist will also search for signs of disease vectors that may indicate a natural cause of death. Pathologists also use powerful microscopes, X-rays, and photography to get the information and documentation they need to determine how an animal died and to make a strong case in court.

Issues that often complicate a cause of death determination in an animal (but should not impact the results of a careful and professional necropsy examination), include,

i. The possibility that the animal may have been struck by additional (non-lethal or crippling) bullets, pellets, or other projectiles days, months, or years prior to the questioned incident.

ii. The possibility that an illicit bow-hunter has shot the animal with a firearm first (because of the difficulty in getting close to an alert animal), and then stuck an arrow into the bullet wound.

iii. The possibility that the animal was killed or fatally weakened by a modern pesticide or poison designed (as the result of environmental protection laws) to decompose rapidly after a few hours of exposure to air or sunlight.

iv. The likelihood that scavengers have destroyed a considerable amount of useful blood, tissue, or bone evidence if the carcass was not found and collected in time. In conducting necropsies involving bullet wounds, it is often extremely important that the veterinary pathologist determine the trajectory of the bullet into or through the body. This information may resolve the question of whether the accused hunter was properly defending him/herself against a charging animal, or illegally hunting a protected species. The information may also be used by investigators, during the interview process, in determining the veracity of suspects and witnesses. Do not testify outside your field of expertise.

Forensic Analysis of Wildlife Crimes

Do's and Don'ts as expert Witness-Pathologist

- Include comments on history, time of receipt of the carcass, carcass post mortem condition, scavenger damage, etc.

- Document the collection of additional trace evidence and establish a "chain of custody" for each new piece of evidence

- As an expert witness portray a professional appearance.

- Answer questions with confidence.

- Consider educational level of jury.

- Use the photographs to document the observations before proceeding to necropsy and provide unbiased scientific evaluation of evidence.

- Preserve the evidence in the original state if possible

- Anticipate as to how to demonstrate to a jury what was found using photographs and/or x-rays.

- Record negative observations in the notes that would exclude alternative explanations by the defense which might be used to advance a different conclusion or throw doubt on the interpretation of the observations. Be sure that the attorney knows the limitations of your expertise, examination results, etc. Do not take personally the defense attorney's attack on your qualifications, examination, judgment, conclusions ethics etc (he is being paid to do!).

- The rights of the defense include: Review your notes, chain of custody, etc. Examine your photo documentation or x-rays. Re-examine all available evidence and split samples of analytical submissions. (Provide unbiased scientific evaluation of evidence and provide the evidence in the original state if possible).

- Before the trial starts, the pathologist should (1). Establish what can and cannot testify based on training and experience (2). Educate the attorney on the science and conclusions based on findings (The pathologist may be asked to develop a set of questions for the attorney to ask so that the answers will provide explanation). Since the trial preparation starts at necropsy table, the pathologist should anticipate as to how to show the jury of what was found and how it pertains to the case using photos and x rays. The pathologist should record negative or normal observations in the raw data to document why other possible explanations were eliminated.

The pathologist should also anticipate the opposite side's explanation and be prepared to counter or support it by the demonstration of the facts.

- Make sure the written report is consistent with any other records, notes or photos on file.

- Use professional quality visual aids to present findings to the judge or jury. Provide unbiased scientific evaluation of evidence. Preserve the evidence in original state if possible

- All photos / radiographs must have evidence identification incorporated into photos

- Valid methods acceptable to the scientific discipline should be used in Analytical procedures.

- Preserve the Instrument calibration logs available.

- Do not give Forensic pathology report or discuss findings with anyone other than submitter or his legal counsel.

- Write report for the court.

- The pathologist should double check notes/photos etc with final report for \ inconsistencies.

- Store the digital images and lock on a CD.

- Secure the radiographs and store them securely.

- Use photos of the radiographs to the court.

2. Role of Morphologist

Morphology is the study of structure and shape. In wildlife forensic science, morphological examinations of submitted evidence items are normally conducted by eye, and with the use of simple, compound optical or scanning electron microscopes. Morphologists can identify the species, and sometimes subspecies of animals ,birds, reptiles using recognizable characteristics of organisms found at crime scenes. They can often determine the age and sex of animals as well as the time-since-death by careful observations of feathers, skulls and skeletons. Identification is not a problem with whole animal. It is very much a problem in the case of wildlife parts and

Forensic Analysis of Wildlife Crimes

products wherein the commonly occurring species-defining characteristics of the animal source may not be present. These identification characteristics typically fall within one of the following morphological categories

- Fur

- Leather and hides

- Bones and skulls

- Teeth, claws, and beaks

- Hooves, horns, and antlers

- Other miscellaneous parts.

Sometimes viscera and bones are removed or scattered and nothing appears to persist. Placing the suspect material in warm water will assist—some components float(feathers), others sink(gizzard stones).Keratinous material such as hair, horn, feathers or scales can provide much valuable information. It may come from the animal itself or from something that it has killed or eaten. Snares, ropes, wires and nets may be attached to the carcass. These should not be removed but submitted with the body. In addition to animal remains there may be other evidence (changes in the soil and vegetations and in invertebrate fauna). These investigations require expert assistance, from forensic geologist, botanist and entomologist respectively. The appearance of animal material will be influenced by environmental factors that cause different types of damage.

Following death, the bones and soft tissues remain relatively intact in situ or they may move (Post depositional positional change) as a result of disturbance by humans and other animals or natural factors like rain, snow, wind, earthquakes. Hurricanes, waves and tides, vegetation growths, falling trees and gravitational effects. One of the most important causes of scatter of bones as well as soft tissues is scavenging by vertebrate or invertebrate animals.

Non-skeletal remains that may provide information in a forensic case are (1) Teeth which are found scattered or buried, (2).Hair, feathers, scales—commonly present in small amounts (dried or caked with soil), even if the carcass has disappeared, (3).Gizzard stones, are commonly found where certain birds (e.g.

galliforms) have died in numbers and readily recognizable by laminated structure, if sectioned, (4).Gallstones and bladder stones are found very occasionally and readily recognizable by laminated structure, if sectioned, (5).Otoliths—It is a feature of some fish and may be present in large numbers where fish have been devoured, (6). Stomach or intestinal contents—these may persist, especially if bone, fur, feather, scales or tough vegetation are present, (7). Pellets (castings).These are usually more compact and drier as stomach contents, and (8). Exoskeletons of invertebrates, including sea urchins are seen in large numbers if any animal has died with large numbers of invertebrates within its gastro-intestinal tract (GIT) Detection of these items requires meticulous sifting of the soil and vegetation followed by laboratory investigation. Placing suspect material in warm water will assist—some components will float like feathers and the others sink like gizzard stones. Forensic entomology is important in animal forensic cases in helping to ascertain how long since an animal has been dead.

Museums hold huge store house of specimens. The specialist staff, are able to offer valuable advice in identification. If Whole live, dead, frozen and taxidermy specimens are to be subject to a postmortem examination prior to identification, it is important that the pathologist does not inadvertently destroy parts of a specimen that would aid identification. A range of items like skeletons, skulls, horns, antlers, tusks (ivory) and teeth seized from species can potentially be identified. Under normal circumstances, with well preserved material, the majority of specimens can be identified to genus, and frequently to species. Most skeleton identifications will need to be undertaken by specialist staff with access to a reference collection. In the event of any legal proceedings, identification will need to be confirmed by an appropriate specialist. A morphological examination is usually sufficient to determine bone and identify hippopotamus, walrus, sperm whale, warthog, mammoth or elephant ivory, but will not differentiate between the Asian and African elephants. Most parts of bird skeletons can be identified to at least family level, even using isolated elements. Major limb bones, skull, pelvis, sternum and pectoral girdle are all suitable subjects, and even incomplete examples of these bones can still yield good results. It may be possible to morphologically identify some parts and animal products like fur made into coats and hats, feathers and parts of birds, products from reptiles such as crocodile handbags, snakeskin wallets, tortoise-shell boxes, turtle shells etc. Similarly it may be possible to identify feathers by looking at size, shape and coloration and making a comparison with reference collections held in museums. Licensed bird ringers are often highly experienced in the identification of feathers from native species. In some cases, particularly with parts of feathers, a microscopic examination may be helpful to detect small variations in the makeup The technique works best on downy material such as contour feathering, since flight feathers often lack the diagnostic structural details. The microscopic structure of these feathers varies between groups of birds, for example pigeon species are different to geese or birds of prey. Even a tiny amount of such feather material can usually be identified to a family level. Whilst this identification is only down to family groups rather than individual species, it can provide useful evidence and may help exclude species which may be lawfully killed or taken. Further identification down to a species level may be possible by using DNA techniques.

As many morphological characteristics are not available it is genetic (molecular biology) section that utilizes the serological proteins and DNA analytical methods (mitochondrial and nuclear DNA) for animal and plant species identification.

Ex: Animal tracks identify the animal behaviour, movement and activity. It also identifies the type of animal species (rear foot bigger in bear), the presence of claws in the track in dogs and absence in cats(retracted claws), tracks further apart indicate the running condition of the animal, size of the animal (adult Vs babies) may be known by measuring the feet, number of feet may indicate the kind of animal at the crime scene—2 feet-birds, 4 feet –dogs, cats, ruminants etc, one can also identify the predators vs prey—the predators have eyes in front/forward (hunts) and in prey eyes are placed on the sides of head (hide). Similarly the skulls and teeth also help in identification of the animal. Antlers are the horny derivatives found in the males of cervidae family. In immature it is found covered by velvet like skin and has blood and nerve supply and hence called as velvet antlers. In mature antlers, the blood supply stops, the velvet covering falls giving solid and rough appearance. These are used as show pieces, buttons, hangers, pistols and butt for medicines. Antlers of 5 major species are found-chital, sambar, swamp deer, barking deer and hog deer. These can be easily differentiated based on colour, roughness and branching patterns in complete form. Claws: Carnivores (Tiger, Jungle cat, Common palm civet, Leopard, Leopard cat, Hyenas, Black bear , Clouded leopard, Himalayan palm civet and Sloth bear) have 5 toes in front paw and 4 toes in the hind paw with sharp claws. It may be refractile or non-refractile.

One new exciting technology to wildlife forensics is retro-engineering. Scans of skulls in a museum can be sent to the laboratory to help them identify a specific skull. This technology is particularly helpful in sorting dog and wolf skulls, which are still posing problems for the scientists.

When an animal evidence is not substantial enough to identify it through morphology, many other techniques can be used.

3. Role of Molecular Biologist (Geneticist)

Many morphological characteristics of the wildlife are not available. The molecular biologists utilize the serological protein and DNA analytical methods, mDNA and Nuclear DNA for animal and plant species identification.

Wildlife DNA forensics is concerned with proving that a crime has been committed through identification of the animal or plant involved. In an investigation the laboratory should be contacted as early as possible. In wildlife crime investigation it is important to know

1. The age of the item,

2. When it was prepared or worked,

3. Where it originated from and

4. Whether it was captive bred, artificially propagated or wild taken.

Molecular biology involves the study of genetic information encoded in the DNA molecule, and the expression of that coding into proteins and related biological structures. Genetics can be particularly useful when the sample is very small or unidentifiable from its morphology. The genetics team uses protein and DNA analysis techniques to identify the family, genus, species, sub-species and gender of a blood or tissue sample and with some species, can also evaluate two different samples in order to determine if they originated from the same individual, also match two tissue samples (from the crime scene and one from the suspect) with statistical certainty using DNA techniques. Wildlife crime scenes should be examined to look for potential sources of human DNA. These may provide a direct link to a known suspect or to an individual already held. In addition, geneticists may be able to provide environmental information about an animal.

Forensic Analysis of Wildlife Crimes

When dealing with specimens or derivatives avoid specific identification unless one is absolutely certain. Native species are usually less of a problem than exotic ones. In some cases, it may not be possible to identify the specimen at the scene. A decision may have to be made whether identification may be possible from notes, photographs and video footage or whether the specimen needs to be physically seized and formally identified at a later time. If specimens are likely to be seized during an enquiry, then suitable arrangements for handling and housing need to be made. However, finding suitable housing for live raptors, plants, parrots or specialist wildlife may be more problematic. Seek appropriate advice at the earliest opportunity. Specimens may have to remain in captivity or under expert supervision for some time until a final decision is made following court proceedings. The use of photographs and video can be used to help corroborate identification.

DNA can deteriorate very quickly if samples are not stored correctly. As a rule of thumb if the sample is dry (e.g. whole trap, fur, ornament, powder) it should be securely stored dry at room temperature, however if the item is wet or damp (e.g. swabs, meat) it should be securely stored frozen. Various methods of DNA analysis are used in wildlife forensics. Polymerase chain reaction (PCR) and DNA profiling are incredibly sensitive and specific methods of identifying animal species and individuals. Mitochondrial DNA is used for determining species and subspecies. Individual profiles often are based on nuclear DNA.

Examples of forensic investigations

- Food: e.g., caviar from Caspian Sea sturgeon, freshwater and marine turtles.

- Pets: many species of exotic amphibians, reptiles and birds.

- Testing to distinguish extant elephant and mammoth ivory products.

- Species identification of cetaceans (dolphin/whale) illegally killed.

- Species determination of birds from eggs.

- Analysis of 'ivory' roulette balls to establish whether they are made of ivory.

- Testing traditional medicine: e.g., rhinoceros horns, bear gallbladders, and various plants.

- Animal furs and skins: e.g., the trade in crocodile and alligator skins; and products sold as souvenirs such as figurines made from illegal ivory or marine turtle shells, or jewellery made of coral.

a. Species Identification

Species' identification is possible for a wide variety of organisms, including mammals, birds, fish, insects, plants, reptiles, and amphibians. It has even been used to distinguish living animals from extinct ones. Species' identification techniques have been used on many kinds of samples, including seized gall bladders, skins and hides intended for the manufacture of clothing, and other items. It can also help determine the species' origin of blood and meat samples from crime scenes. It is the crucial evidence to support an investigation where illegal trade in endangered species is suspected but products of protected and unprotected species can look similar. More than one method will be used. These include protein analysis, chromatography, immuno-diffusion, DNA sequence analysis, and mass spectrometry. The forensic process of determining the species origin of an unknown tissue normally begins with a series of screening (immunological) tests designed to narrow the possibilities down to the species comprising a single family (e.g., bears - family Ursidae, or deer - family Cervidae). Once the family source of the specimen is determined, the examiners can go forward with either protein or DNA polymerase chain reaction (PCR) analysis (along with the necessary and comprehensive databases) to determine the actual genus and species involved.

A quick way to identify a species is by testing the hemoglobin in tissue or blood samples. The test looks at gene markers of the hemoglobin chains (takes only five or ten minutes). Using an instrument the matrix-assisted laser desorption ionization (MALDI) time-of-flight mass spectrometer. The MALDI technique must be verified by another test, usually DNA analysis (DNA can be extracted from the mitochondria, then sequenced and compared to species in a gene bank. This is the classic "DNA fingerprinting" technique used in some human forensic cases. DNA fingerprinting is available for wolves, elk, moose, mule deer, and bears (just black bears and polar bears).

Electrophoretic techniques were used to prove illegal hunting of a protected species as small amounts of dried blood stains could be clearly identified as of particular species of animals (Example--otter origin). Species differentiation between the otter (Lutra) and pheasant (Phasianus colchicus) can be made based mainly

Forensic Analysis of Wildlife Crimes

on small differences in electrophoretic mobilities of the respective albumins and haemoglobins in one-dimensional separations. Additional differences in minor proteins of the two species can be obtained by two-dimensional electrophoresis. A range of methods—Morphometry, chemical tests, DNA profiling, meat analysis, Mass spectrometry, Thin layer chromatography and X-ray fluorescence are used to identify a range of products including musk, ivory, bear bile, ginseng, whale oil, caviar roe, and rhinoceros horn. They may also provide information on geographic origin, or to assess captive breeding claims.

Unlike human DNA fingerprinting, in wildlife cases, sample sizes are much smaller particularly small for endangered species. With endangered populations of fewer than 4,000 individuals, it is difficult to know which DNA characteristics are "standard" and which are unique to that individual. When some members of a small population are in zoos, it becomes even more challenging. "Zoo samples are not representative of populations (in the wild)". Basically, the more endangered, the less likely an individual can be distinguished from the rest of the population.

b. Gender Identification

Physical examination of the size and shape of bones, can reveal the age, or at least age class (less than a year, yearling, adult), and the gender. A variety of empirical or scientific tests can be used for identifying the gender (sex) of a sample, as well as determining if two or more samples originated from the same individual animal or how many different animals are represented in a collection of samples. A number of nuclear DNA-based gender-determining tests are available for blood and tissue samples from mammalian species. DNA testing can also be used to determine whether a parental or filial relationship exists with another sample, and also the minimum number of animals found in a collection of samples. The tests generally use PCR amplification to detect specific sequences of the ZFY and/or SRY genes, both of which are located on the mammalian Y chromosome.

c. Individual Identification

Early work on individualizing animal blood and tissue samples involved multi-locus DNA probe hybridization techniques. To go from species identification to identifying an individual animal, short tandem repeats (STR) is used. Within the DNA strands of each animal in a particular species, there are large sections that are identical and rarely change. There are also short, repeated sequences interspersed within these more stable sections that vary from individual to individual. Some individuals may repeat these sequences only four times. Others may contain 22 repeats of the sequence. By comparing these sections from two or more samples, one can tell whether or not the samples are from the same individual animal. For STR tests, DNA is extracted from a whole cell or cells rather than extracting DNA solely from the mitochondria.

The minimum number of animals found at a scene can be determined by counting the parts of animals left at a scene. As an example, the number of legs found at the investigation site can be counted or the number of animals can be estimated by examining the amount of meat found. Matching of samples can sometimes be accomplished by simply placing two parts together physically, like pieces of a puzzle. This is a relatively rapid and inexpensive procedure and is very successful in court. However, if this is not possible, DNA "fingerprinting" can be used. New developments in DNA testing have allowed an increasing number of sample types to be analyzed, including partially degraded samples and those with only small amounts of DNA.

d. Population Identification

It is sometimes possible to determine if an animal originated from one specific geographic area as opposed to another. Food animals eat is affected by the minerals in a given area. Certain minerals will be more common in one area than another, and this leads to differences in detectable levels of them. For example, the location of harvest sites of freshwater mussels can be determined by trace metal analysis of shell samples. The natal areas of migratory birds and the geographic origin of African ivory can also be determined by this type of analysis. Fish raised in commercial fish farms are often fed a diet high in a fatty acid called linoleic acid. In contrast, this compound is found only rarely in nature. By measuring linoleic acid levels, it can be determined whether a seized fish was farm-raised or was from the wild.

DNA analysis can also reveal the source population of an animal through genotype studies. Strict chain of custody procedures must be followed at all times.

Collection of material for investigation

When collecting material for genetic analysis, the following should be kept in mind.

- The condition of the material at the scene (dry, frozen, decomposed, etc.).

- Amount of material available, Personal protection and

- The investigative question (species identification or individual matching).

- Only a small amount of (ideally) dried tissue is needed for DNA analysis.

- Contact the forensics laboratory or other testing facility to discuss specific packaging and shipping issues before shipping the item(s) for analysis. Never package bloody clothing in plastic bags (use paper bags).

Materials suitable for genetic analysis include fresh tissue, gut piles, bones, horns, antlers, hair (with follicles),teeth ,claws and nails, feathers (less DNA available in downy feathers),skins (dried, salted, or tanned),blood (fresh or dried), body fluids, cooked meat, crafted items with animal parts, some medicinal items and gall bladders. If the tissue at the scene is fresh, wet, or frozen (i.e.: a carcass, a gut pile, or fresh/wet blood) the sampled material should be kept frozen during storage and shipment. Sample the best quality tissue available. If possible, avoid sampling from connective tissue and cartilage. If the material at the scene is dry [i.e.: blood on a substrate (rock, leaf, trap, etc.), dried tissue, bones, antlers, horn or hair], the sampled material should be kept dry for storage and shipment. If possible, scrape the blood or tissue into a paper envelope without using water. Once the material is sampled, the evidence package should be properly sealed and initialed. Material shipped frozen to a forensic laboratory or other testing facility should be packaged with blue ice or a similar cold pack, and shipped overnight. Material shipped dry should be packaged securely to maintain the integrity of the evidence container.

4. Role of a Criminalist

A criminalist is an expert in the scientific study and evaluation of physical evidence in the commission of crimes. The criminalist employs forensic techniques that detect gunshot residues and map a bullet's trajectory through flesh. Criminalists are responsible for analyzing other types of evidence that might be relevant to a wildlife crime. They match firearms to bullets, tyres to tyre tracks, shoes to footprints, paint chips to vehicles, fibers to clothing, fingerprints to a person, and poisons to a bottle in a suspect's garage. To match a bullet found in an animal to a bullet shot from a particular gun, striations on the bullet's surface are looked at through a microscope. Microscopes are also used to see if a fiber is natural or synthetic. While the other sections of the laboratory deal mostly with animals, the criminologists tackle the bits and pieces of the crime scene that link a person to the criminal act.

The criminalistics team conducts 'standard' police crime laboratory examinations on firearms, bullets, cartridge casings, footprints, shoe and tyre impressions, tyre tracks, paint, glass, fibers, and items bearing latent fingerprints. Instruments utilized include a scanning electron microscope, and a Fourier Transform Infrared (FTIR) microscope. The primary responsibilities are visual and microscopic comparisons of evidence. Criminalist also determine the caliber, possible source of weapon and individualize the cartridge cases and bullets.

Wildlife forensic scientists frequently process evidence from an illegal hunt. The events associated with a typical illegal hunt often involve the following categories of evidence.

Trace evidence

- Firearms

- Other weapons

- Impression marks

- Latent fingerprints

- Questioned documents

Firearms are regularly used in poaching cases, the illegal killing of birds of prey, and may be used to dispatch illegally trapped or snared animals The matching of recovered shotgun cartridges, other ammunition cases and air weapon pellets/ projectiles to particular weapons can provide important evidence. A firearms examiner may be able to provide useful information about types of ammunition or weapons used, the range at which a weapon was discharged etc. When shotgun cartridges are recovered from a crime scene, attempts should be made to retrieve the wadding which is likely to be in the vicinity. Bullets and air gun ammunition may be recovered from the bodies of victims for comparison with weapons held by a suspect. Care must be taken not to mark these items, by the use of metal instruments during a postmortem examination. Instrument marks introduced to a bullet could prevent a conclusive identification.

Forensic Analysis of Wildlife Crimes

Bullets or bullet fragments or shell casings are often found at crime scenes and are a valuable form of physical evidence. They may also be collected from carcasses of animals removed from the kill site by hunters. Metal detectors are used to help find bullets and bullet fragments, which are then analyzed in the laboratory. The make, type, and caliber of firearm can often be determined and perhaps even matched to an individual weapon. However, the success of such analysis depends on the condition of samples.

The typical animal kill involves a high-powered (and large-caliber) rifle or a shotgun at relatively long distances (50-300 m). It is often difficult for an illicit hunter to retrieve expended casings due to the nature of the typical hunting area (brush, trees, and ground cover). It is extremely difficult for a crime-scene investigator to locate the shooting point, much less the expended casings. The suspect may take the victim (as a trophy or meat) back home. Hence the bullet is likely to either be in the carcass of the animal, or in the "gut pile" left at the scene (which can be matched to the trophy head or meat with DNA techniques). The typical illicit hunter rarely discards the weapon after a single illicit kill since he spends a lot of money on a rifle or shotgun. Thus, it is very likely that a succession of illegal kills can be linked to a single poacher by matching the spent bullets or casings to the rifle or shotgun. In some circumstances, it may still be possible to recover bullets, which may have passed through an animal and perhaps embedded themselves in the ground or in a tree. A metal detector may be useful for this purpose. Embedded bullets should be removed without causing damage.

Other weapons such as crossbows and catapults may be used to commit wildlife crime. Examination of these types of weapon will need to be done by forensic service providers. When dealing with specimens or derivatives it may be advisable to avoid specific identification unless absolutely certain. Other hunting weapons typically associated with an illicit animal kill include long bows and arrows, crossbows and "bolts", spears, spring traps, poison discharge devices and nets.

Collection of Materials for Investigation

Firearms

Make sure that all firearm evidence items are unloaded. Do not scratch initials or item numbers into the stock or receiver of firearms. Record the make and serial number and attach a tag with the item number. Do not shoot black powder firearms to unload them. Remove the cap/primer/flint and have a person with the proper expertise in handling such weapons to unload them by pulling the projectile. Place firearms in a padded plastic case for shipping. Cardboard boxes are not sufficient for protection, as the firearms may shift and break out of the boxes in transit and stocks and scopes may be damaged. Leave unloaded detachable magazines in or with the firearms.

Bullet and Cartridge Casings

Gently wash the blood off of bullets and then air dry if DNA processing is not needed. **Package the bullets in paper envelopes.** Cartridge casings often pick up latent fingerprints when being loaded into rifle or pistol magazines. Do not store bullets in plastic bags or plastic or glass vials, since any moisture, combined with blood and tissue, will putrefy and form acids that will etch the surface of the bullets and destroy fine detail needed for comparison purposes. Do not use pliers or forceps to remove bullets, as these instruments can scratch the surface that is important for microscopic comparisons. Do not scratch marks on bullets or cartridge cases. Mark the outside of the container with the item number and initials. Do not add your own prints, or smudge the ones already there, when you collect and ship the casings. Dry wet cartridge cases or shot-shells, and place them in paper containers. Always consider the possibility of latent prints, and preserve/protect the items accordingly.

Impression Marks

Most illicit hunting situations occur in remote areas. The suspect will leave tyre tracks and boot impressions in soil, mud, or snow and the fact that these tyres and boots are typically used in off road situations makes it all the more likely that the tyre or boot treads will possess individualistic wear marks.

Vehicles, tools and instruments feature in a range of wildlife offences. Examples could include: (1). Footwear marks at a badger digging incident, (2). Climbing iron marks left on a tree by an egg collector, (3). Examination of close rings fitted to a bird which may have been tampered with, (4). Examination of quad bike tracks at a wildlife poisoning scene, (5). Comparison of pliers seized from a suspect with the cut ends of razor wire put around a tree to protect rare nesting birds and (6). Where knives or instruments have been used to kill or injure animals and have come into contact with bone.

Impressions left on a variety of surfaces by shoes or tyres can be used for identification purposes. The make of shoe or tyre can be identified and sometimes, if there are unique marks, cuts, or signs of wear present on the exhibit, a match between a specific exhibit and the impression can be determined. Reference collections of the makes of shoes and tyres are used for comparison. Evidence is usually collected at the scene by the investigating officer and comparisons are done by qualified analysts in a laboratory setting. Impressions can be photographed or plaster casts prepared. Dental stone, reinforced plaster and sulfur are some of substances that can be used to cast the prints left in soft surfaces such as mud or snow after photography. Casts are preferred to photographs alone, since a three-dimensional cast yields more information. Talcum powder can be used as release agent between the print and casting material. Snow prints can be hardened and protected from the casting material heat with precasting material such as Snow Print Wax. Hair spray or spray polyurethane can be used to stabilize dust or sand prints. Casts of snow impressions can be made if first treated with a special spray wax prior to casting. Photographs should be taken before and after casting. It is essential that some sort of scale, such as a ruler, be used properly and included in photographs. Imprints made on dusty surfaces can be photographed and sometimes lifted, using a technique called electrostatic dust lifting. Bloody footprints can be stained and photographed. Collecting footprint or tyre impressions must be done as soon as possible so that the evidence is not lost. Under water prints can be cast by sifting casting material over the prints.

Friction Ridge Evidence (Finger, Palm, and Bare Foot Prints)

Friction ridge skin – raised layers of skin with openings for sweat glands – covers the palms of the hands and soles of the feet. During fetal development, these ridges form patterns that remain unchanged. The stable and complex characteristics of friction ridges enable a form of identification that law enforcement has used for over a century.

This type of evidence can yield important information for the investigator if properly collected, packaged, and transported. Prior to collection the investigator should determine, early in the scene and evidence review process, what items would be suitable for processing. That decision should be made to assure the best possible chance to develop friction ridge detail. The surfaces of items can be divided into porous and non-porous with the nonporous being the most fragile. This is because the residues are on the surface as opposed to being absorbed into porous materials.

With this being the case, the way the item is collected, packaged, transported, shipped and stored, will have a significant impact on the successful development of the latent print. Do not add your fingerprints to latent prints that may already be on the items, and package the items in such a way that they do not rub together. Evidence items with nonporous surfaces must be handled carefully so that the surface areas to be processed are not touched (even with gloved hands). The investigator must also consider the possibility that other informative materials (i.e.: blood, a print in blood, saliva, trace evidence, or a tool mark, etc.) may be present on the same surface area. Depending on the materials present, this may change the sequence of processing the surface to avoid damaging or destroying useful evidence.

Important note: Any information that is known about the item's history (such as weather, fire or chemical impacts) should be submitted with the item.

Latent Fingerprints

Latent fingerprints are the classical means of linking suspect, victim, and crime scene through physical evidence. The following types of latent-bearing evidence are frequently submitted to a crime laboratory in wildlife cases.

1) Firearms

2) Expended casings (shotgun and rifle)

3) Knives

4) "No trespassing" and "no hunting" signs

5) Game tags and

6) Import / export permits

Fingerprints

The use of fingerprints to identify individuals is a commonly used forensic technique. Suitable surfaces at crime scenes, or on items potentially handled by a suspect may be appropriate for this method. Items at outdoor scenes may need some protection from the effects of inclement weather prior to examination for finger prints. Fibres may be transferred from clothing or items used in the commission of an offence. Hair and fur comparison could include both human and animal samples.

Evidence suspected of bearing fingerprints can be submitted for analysis. Usable fingerprints can be found on a variety of surfaces, including paper (for example, on documents accompanying shipped items), wood, metal, and plastic. One very useful source of fingerprints is the under surface of adhesive tape used to seal packages. Although it is not always possible to get clear, complete prints, even partial prints may be useful. In addition to fingerprints taken from actual suspects, various databases of known fingerprints exist for comparison purposes. Fingerprint evidence has a long case history, especially for non-wildlife crimes, and has proven valuable in court many times.

Trace Evidence

Place paint chips in a marked paper-fold and then place the paper-fold in an envelope with the item number and initials on the paper-fold and the outside container. Paint chips placed in a standard envelope tend to break up and sift out the corners. Hairs and fibres can be placed in a marked paper-fold. Avoid using tape to collect fibres and hairs as they may be very difficult to remove from the tape glue during laboratory analysis.

Examining Documents of Questionable Origin

Examination of documents can confirm whether they have been counterfeited, forged or falsified. Handwriting comparisons can connect documents to suspects. The repetitive manner in which documents have been altered can provide strong evidence for a singular origin or suspect. Document examination can also indicate links to other shipments of goods and can provide evidence against a suspect for other offenses.

Documents of questionable origin are frequently encountered in wildlife investigations involving the import and /or export of wildlife parts and products. This is typically in relation to documents that have been forged or altered to try and disguise the provenance of illegally held items. The question most frequently asked by the investigator is whether or not the seized documents (typically import/export permits) are valid. This can be an extremely difficult question to answer when the authorizing seals vary between countries, the names of individuals authorized to approve import/export documents change on a frequent basis, and the shipments must be cleared (the documents examined and approved) at the local port of entry.

The following documents are typically associated with a wildlife case.
Forged or altered hunting licenses.
Forged or altered game tags and
Forged or altered import/export permits.

Digital forensics /Computer forensics (Forensic digital evidence)

The examination of digitally stored records and information is a rapidly developing area of forensic examination. The different locations where digital evidence may be found may include:

- Locally on an end user device – typically a computer, mobile, smart phone, digital camera, satellite navigation system, USB drives and portable storage devices.

- On a remote resource that is accessible to the public, for example websites used for social networking, discussion forums, and newsgroups.

- On a remote resource that is private, for example Internet Service Providers logs of user activities, mobile phone records, webmail accounts and remote file storage.

There is a wide range of places where electronic information may be stored, and which can hold huge quantities of information. Information from these sources can provide important evidence and intelligence information. Check the diaries, notebooks and scraps of paper, which may contain passwords. The integrity of computer evidence needs careful attention. Any activity on a computer leaves a trace and files with dates after the seizure date may render the evidence inadmissible in court. **Potential evidence items include** E-Mail, User Documents (Letters, memos), **Internet browsing:** Favourites, Temporary files, History, activity logs, Financial Records---Price lists, Vendors and Buyers, Videos, Graphics (pictures), Registry Information, Documents downloaded, E-mail addresses, Screen Names, IP address, Customer databases, Scanned images of key documents, Chat Logs and Programs being used.

5. Role of an Analytical Chemist

Chemists may be asked to identify poisons and pesticides, characterize the contents of Asian medicines, and provide species identification, when possible. Chemical analysis techniques are used in a wildlife crime laboratory to provide toxicological

Forensic Analysis of Wildlife Crimes

information for a veterinary pathologist conducting a necropsy. Analytical chemistry procedures employing chemical biomarkers are used to identify the species source of animal products such as bear bile and deer musk. Standard toxicological methods are routinely employed to identify chemical baits and poisons used to trap and kill wildlife. The chemist uses a wide array of scientific instruments (gas chromatographs (GC's), a gas chromatograph/mass spectrograph (GC/MS), a liquid chromatograph/mass spectrograph (LC/MS) and a high pressure liquid chromatograph (HPLC)) to identify pesticides, poisons, general chemical compounds, and traditional Asian medicines. Fourier transform infrared spectroscopy (FTIR), is used for different aspects of chemistry including the analyses of bear and shark gall bladders for bile and also turtle shell analyses. Diffuse Reflection Infrared Spectroscopy (DRIFS) analyses are used to detect keratin in turtle shells.

When COD or species identification cannot be determined using biological or analytical approaches, the chemistry unit uses blood and tissue evidence or any derivatives products to examine the chemical and molecular structures for species identification and COD. The chemist also provides toxicological methodologies to identify toxins or poisons useful in determining the manner of death.

Collection of Material for Investigation

Pesticides and Poisons: Pesticides and poisons are often used to deal with 'pest' animals, or using a bait animal or a food product laced with the chemicals to kill a difficult –to find predator, the latter situation often involving a 'bait'.

Avoid using glass containers to package and ship pesticide and/or poison evidence items, as the containers may break during transit.

When investigating poisoning, consider

- Agricultural practices in the area
- History of previous poisoning incidents
- Local pest control activities
- Locality and distribution

- Epidemiological profile
- Clinical signs
- Variety of species
- Predator/prey relationships and
- Who is authorized to use such products in that area.

The crime scene investigator should consider

- The source of the poison
- Ingestion
- Contact
- If the victim was the primary or secondary target
- Primary consumer of poison bait
- Consumed another animal which was primary target of poison.

Documenting the poison route involves

- The sequence of ingestion
- Identification of food items
- Analysis of individual items
- Consideration of water versus cutaneous (absorbed through the skin) route and
- Determining time delay and travel from the potential source of poison

Personal protection

- Always wear protective equipment (gloves, mask, coverall suits to go over clothing, booties, etc.).
- If you can smell "chemicals", your pesticide exposure may be unsafe.
- Do not transport baited carcasses inside a vehicle or in the trunk.

Forensic Analysis of Wildlife Crimes

- Do not contaminate camera, writing utensils, or other equipment.

- Double glove so that the outer contaminated glove can be removed if it develops a hole.

Pesticide and poison collection kits are available from forensic laboratories that will conduct the analysis of the collected samples.

Evidence Collection and Submission

Forensic analysis is considered to be legally valid only when sample integrity and evidence continuity are safeguarded. Accurate analysis of the sample collected is impossible if it becomes degraded or contaminated. There are a number of ways this can happen.

1. Cross contamination of samples during collection: This can be avoided by using clean knives to cut the samples from a carcass, between samples.

2. Degradation of samples due to improper storage and shipment techniques: Once evidence is collected, it must be stored under conditions that prevent further degradation.

3. Cross contamination of samples during shipment of evidence: Forensic samples are often shipped in plastic bags. Although this is acceptable for wet items as long as they are kept frozen or cold, samples with sharp edges, such as knives, pieces of bone, and rocks, may pierce the packaging during transfer. Even pieces of tissue, which do not appear sharp when fresh, are very hard when frozen and can wear holes in packages if allowed to bang against other packages during shipment. Bags that appear intact when the contents are frozen can be leaky by the time they arrive at their destination. Dry items shipped in paper containers such as bags or envelopes are subject to the same problems of piercing and mixing. Some of the techniques described, especially time of death measurements or tyre and shoe impressions, must be performed in the field.

Considerations for proper shipping

Evidence must be properly sealed and shipped in appropriate containers. If the items need to be kept cold, insulated containers such as coolers, with ice packs, must be used. A cardboard box containing the evidence items wrapped in newspaper is not suitable. Make sure that whatever is used as the source of cold cannot leak and damage the evidence. Often evidence is shipped to laboratories via a courier service. Even though these services 'guarantee' shipment within a certain time period, it is wise to prepare for a period longer than expected. If the officer retains part of the evidence, this can be used in the event of loss or damage to the original shipment. In addition, evidence is available if the defense counsel desires to have it tested by an independent laboratory. Proper documents must accompany the evidence. This applies to evidence crossing international borders that is subject to inspection by customs officials as well as evidence involving species whose movement is restricted by international agreements or other laws. Take all possible steps to avoid loss of evidence continuity. Have all necessary permits in advance. Officers should contact the laboratories prior to shipping any evidence. Laboratory personnel will be able to discuss the laboratory's ability to do the required analysis, predict the time it will take, estimate costs, and give advice on the best method of shipment. Laboratories should also be available to answer questions regarding the collection of evidence. Agree on the method of shipment and confirm the approximate arrival date. This can enable the receiving laboratory to trace lost shipments before it is too late.

Legal requirements for prosecution

Evidence submitted for forensic testing is subject to a number of requirements.

Chain of custody (evidence continuity) and evidence preservation - Failure to maintain a proper chain of custody will result in inadmissible evidence. If evidence is not collected and maintained in a proper manner, it is possible that it will not be usable for forensic analysis. **Evidence handling procedures, at all stages of an investigation, may be questioned in court. Forensic analysis does not replace proper investigative technique** and will not save an investigation if proper procedures are not carried out. Wildlife officers and laboratories share responsibility in getting proper evidence to court. It is the officer's responsibility to ensure that evidence collection and other portions of the investigation are done properly. It is the forensic laboratory's responsibility to have properly-trained laboratory personnel performing the analyses, since it is essential that analysts be able to qualify as expert witnesses should the case go to court. **Forensic laboratories must handle and store evidence properly**, maintaining the chain of custody once the laboratory has received the evidence.

Forensic Analysis of Wildlife Crimes

6. Role of the Forensic Specialist /Investigator

1. The primary role of a forensic specialist is to provide the information necessary to make decisions regarding a case by the law enforcement officials, judges, and jury members (decision to prosecute, render a verdict, etc.).

2. An investigator should conduct the examination of the animal in an unbiased manner (to avoid unnecessary influence in the interpretation of findings). He has a great responsibility of maintaining the admissibility of evidence and providing accurate documentation of evidence and findings (failure to thoroughly document the possession and transfer of evidence may result in the loss of vital information due to inadmissibility). The investigator has to give opinion or interpretation based on the available evidence which can withstand the scrutiny in a court of law.

3. The task of investigation begins with the submission of a specimen(s) by law enforcement officials.

4. The forensic pathologist should maintain secure control over specimens or portions of specimens and collected information until their return to law enforcement officials or disposal.

5. The investigator should render an informed opinion or interpretation based on available evidence, his or her professional training, and prior experience. For successful presentation of evidence and findings, all of these actions and interpretations must withstand scrutiny in a court of law.

Chain of Evidence

In any forensic necropsy it is very important to maintain continuous control over all specimens and information which enables one to testify with certainty as to their identity. This control is referred to as maintaining a chain of evidence or chain of custody. The chain of custody means "that every individual who has sole custody of the evidence must appear to testify that the evidence gathered in the field is the same exact evidence examined in the laboratory". It also includes secure storage of evidence, photo documentation, maintaining a proper written record of findings, and documenting transfer to other professionals for necessary ancillary tests. Anyone that comes into possession of evidence may be called into court to testify about a specimen's identity.

1. The chain of evidence begins with the enforcement officer who first encountered the animal(s) in the field. He should ensure that the evidence is admissible in the court of law.

2. It is the responsibility of the enforcement officer to maintain the custody over the animal until it reaches the laboratory when it becomes the pathologist's responsibility.

3. The Enforcement officer (EO) should bring the specimens to the laboratory and turn over directly to the pathologist who will perform the necropsy whenever this is possible. Specimens can be submitted through a receiving clerk or technician rather than directly to the pathologist but this may complicate the chain of custody. When specimens are submitted for forensic necropsy, the officer maintains a chain of custody log. The person receiving the specimen will sign and date the log, thus documenting that he/she received a specific piece of evidence. It may be easier if the person receiving the evidence is the forensic pathologist responsible for the case. Remember, any other people that come into possession of the specimen are additional links in the chain, and additional opportunities for mistakes to occur. Specimens for forensic examination should never be left at a laboratory without being received by a responsible individual, lest it may break the chain of custody.

4. If any specimens are to be maintained by the forensic pathologist, they must be kept in a secure area to avoid risk of contamination, mix-ups, or tampering. Specimens must be kept in a secure manner and transfers of custody within the laboratory must be recorded, as each individual involved may be required to testify. In the laboratory each specimen must be clearly identified with well attached labels or tags at time of submission. There must be strict attention to maintaining a chain of custody when specimens are received from the field and in all subsequent procedures, so that the identity of specimens and information resulting from the necropsy is beyond question. Each specimen must be clearly identified with well attached labels or tags at the time of submission.

5. The pathologist receiving the specimen from the officer should note the time of receipt on the evidence tag or chain of custody form attached to the animal(s), and then sign the tag in the presence of the submitting officer. This tag remains with the animal until time necropsy is performed.

Forensic Analysis of Wildlife Crimes

The pathologist and enforcement officer must discuss and agree on what can and what cannot be accomplished by a necropsy and the tests that the laboratory is equipped to provide, what specimens need to be retained for potential use as evidence, how these specimens will be maintained and returned to the enforcement officer, and how and when one should dispose of portions that are not required. Normally, physical evidence that may be used in court is returned to the enforcement officer who signs for receipt of any specimens when they are returned to his or her custody. In some cases, it may be advantageous for the enforcement officer to attend the necropsy and collect specimens, such as bullets or suspected poisons, directly.

Scene Recording

Photo documentation is an important part in any crime investigation. Photographs and sketch plans of the scene are an invaluable record of what was observed and are useful in court for explaining the nature of lesions. Video footage can also be considered. This is best undertaken by crime scene investigation (CSI) officers who are experienced with this type of scene recording. Where CSI officers are unavailable, efforts to prepare sketch plans and take good quality images should be still made at the outset of the investigation and before any exhibits are seized or the scene disturbed. Next series of photographs documents the actual evidence items as indicated by the play cards. Each play card will indicate the evidence items and the photographer should take 3 photographs of each. The first photograph shows the evidence item next to the numbered play card, the second is a close up of the evidence item with just a portion of play card and the last photograph is a close up of the evidence item(in which just the item and the adjacent ruler are visible). **At a crime scene consider a range of photographs: Wider range** shots showing the surrounding landscapes. These provide information about the type of land use and who may potentially be involved. **Middle range** shots showing the immediate environment of the crime scene. Where items of interest are not easy to see on wider shots consider marking their location with something more obvious. CSI have numbered markers or small flags. **Close ups** show small items or specific points of interest. This could include footwear impressions, animal prints, discarded items, injuries to wildlife casualties, hair, blood, feathers etc. These should incorporate a scale ruler or item of known size. Ensure time and date are accurately set and all photographs and video material should be retained, even if not intended for use in court. No images or file names should be deleted or altered in any way. The last image of the crime scene should be followed by a photograph of the play card (including the time of departure) used to initiate the photo-shoot. Whole range of photographs should be taken even if some appear to be of little or no relevance at the time. If more than 1 person is involved in the photography, record who took each picture (since the photographer may need to be identified during court proceedings). Avoid other people to take photos at the crime scene because they may be asked to testify. Do not delete any photograph from the crime scene series since the numerical continuity of the digital files is important than the concerns over the quality of the photos. The investigator should transfer the digital photographs from camera and archived. Even if photographs are taken, sketch maps should still be considered to highlight the location of points of interest. The location of an incident is often very important, particularly as the scene may need to be revisited at a later date. A GPS device can be extremely useful in precisely pinpointing a location. Where these are not available, consideration should be given to measuring or pacing from known fixed points, three if possible, to be able to triangulate the precise offence locus at a later date. The use of compass bearings to selected fixed points can also be helpful.

While doing necropsy, document the identity of the specimen with photographs. Photograph of intact animal along with the identification tag before the commencement of necropsy is taken. Throughout the entire necropsy, photograph significant findings (depicting the lesions and areas of interest) with marker. Photos provide visual support as to the identity of the specimen and link the evidence number used by investigators and the accession number used by the laboratory for histology, ancillary tests, etc. Photographs are usually prepared as large format color prints. The pathologist himself should be able to take photographs. If not possible she / he should be present when they are taken since it reduces the complications as to the identity of specimens shown when the photographs are used in the court of law. Every photograph should contain information that clearly identifies the animal, the date, a size reference scale and the identity of the pathologist. Cards (matted paper) of different colours (grey, green or brown) with spaces for reference scale and relevant information are prepared before the necropsy. Markers and symbols (numbers and letters) are placed directly on the specimen to indicate the areas of interest. Pointers can also be used. Each photograph should be accompanied by a clear description of what is illustrated (at the back of the photograph) and should be prepared by the pathologist himself before giving to the enforcement officer.

Tag may be signed by the pathologist in the presence of the submitting officer. The evidence tag remains with the specimen until the necropsy is completed. It contains the accession number for the laboratory, the signature of the pathologist, the date and a scale. Markers insure that every photograph has an inherent, proven link to a specific case. If any specimens are to be maintained, they must be kept in a secure area to avoid risk of contamination, mix-ups, or tampering.

When the animal is submitted for necropsy, the veterinarian should maintain a number of records. During the course of the necropsy the veterinarian / pathologist must have the foresight to collect the appropriate samples in the correct manner, avoiding cross contamination In addition, the samples would have to be stored in proper containers and shipped in a secure manner to the appropriate laboratory. All ancillary tests must be performed using methods that are the current standard for a specific field and defensible in court. Another important consideration in wild animal vetero-legal necropsy is the use of experts in other disciplines to perform necessary analyses. For example, conclusions about ballistics or projectile character would be better assessed by a forensic firearm specialist and not a pathologist.

4 | Forensic Necropsy

Postmortem Examination (PME) of a dead wild animal or its parts is an important component of wildlife forensics. Necropsy methods do not differ substantially from species to species in terms of the need for standard protocols, appropriate equipment and proper record keeping. The dissection should be carried out with sensitivity and care. Consultation with a pathologist, toxicologist, forensic veterinarian, or other forensic professionals may be beneficial during necropsy, evidence collection, and/ or interpretation of findings.

The principal purpose of forensic necropsy (FN) is to determine the cause and nature of death (cause of ill-health, underlying abnormalities / pathology) of the animal being examined. Identification of the animal as to species, sex (reproductive status) and age is important for enforcement purposes and the pathologist may need to seek assistance in these matters. Determination of time sequence in which events occurred, the general health of the animal and the presence of preexisting conditions that may have influenced its death may also be important. It is necessary to collect information and perform ancillary tests to rule out alternative explanations, to show that the animal was in good health and was not suffering from an infectious disease at the time of death.

Examination of the carcass of an animal (necropsy) can reveal much useful

Forensic Necropsy

153

information. The cause and manner of death can often be determined. If poisoning is suspected,-- samples can be taken during necropsy and sent to a laboratory for testing. If wastage or abandonment of the carcass is suspected, - bacteriological testing can be done to determine if the meat is spoiled. Necropsy can be used to determine if an animal was killed with a bullet or an arrow, by collision with a vehicle, by trapping, or by predators. The trajectory of the projectile or its direction of entry can be ascertained and can be helpful in confirming a suspect's story (i.e., was the animal shot in self-defense or not?). X-rays can be used to find whole bullets or fragments, which can then be tested by ballistics experts. Stomach contents can shed light on a number of situations. They may reveal if an animal has been causing crop damage, if it has been poisoned, or, in cases where the animal is suspected of an attack on humans or domestic animals, if parts of a victim are present. Bite mark analysis refers to the examination of victims (either human or animal) of predator attacks, to determine what predator species was involved. Bite mark examination can also be used to determine if a victim was attacked or if an already dead individual was scavenged. This is important, for example, if it is necessary to decide whether compensation should be paid to a livestock owner for a dead animal. It may be important to show that the correct animal was destroyed. DNA from the attacking animal can be isolated from the saliva deposited at bite wounds and tested to see if it matches DNA from a suspect animal. Animal hairs and other evidence may need to be collected from the victim as well, for matching purposes. Because of the sensitive nature of predator attacks, it is best to have sample-handling protocols in place ahead of time to avoid confusion and possible complications through disputes between different jurisdictions.

Field necropsies need special preparations. Many wildlife necropsies are performed in PM room, away from where the animal died. A site visit is always advisable. Do not investigate the carcass until the area where it was found has been properly searched and relevant people questioned. Not all the Wildlife forensic necropsies are the task of specialist veterinary pathologist. In some countries may be performed by biologists from diverse disciplines without training in PM techniques and interpretation of findings.

In wildlife work, in extensive game reserve, the person who carries out initial examination of the dead animal is a game ranger or warden, who finds the animal first not a biologist not a veterinarian or not a pathologist. The PM record should bear the names of such assistants since they may be required to be identified or to give evidence in court. Adequate assistance is usually essential to provide practical help (technicians, attendants) and if necessary to offer corroboration of findings. The amount and type of assistance needed will depend on species. Rhinos require more people to do PME than snakes (need knowledgeable people who are not averse to handle poisonous snakes).All those participating in the forensic necropsy must be properly briefed beforehand and reminded that a legal case may result.

PME can yield data on morphometrics, organ weights and organ/body weight ratios and on gross and histological appearances of tissues. In cases of rare or endangered species, when time is short , a request may be made to a zoologist with the knowledge of the species that is being examined to attend the necropsy to advise on special morphological features and to record biological (non pathological) data or may consult a suitable qualified person before doing PME in order to ascertain zoological data, if possible. It should be done by a specialist veterinary pathologist not even by a veterinary practitioner.

Wild animals are presented often as decomposed carcasses or parts. The rapidity with which decomposition depends on the method of wrapping, depth of burial, ambient temperature, soil type and the presence or absence of water. Wild animal pathologist may also come across wild animal that has been fixed in formalin, glutaraldehyde or an alcoholic preparations. It is important to note that many species of wild animals predate and devour bones, possibly in order to obtain dietary calcium. On receipt of carcass, strict hygienic precautions should be followed while handling and unpacking carcasses regardless of history, because of possible transfer of pathogenic organism or poisons.

Identify the hazards and assess the risk for severity and out come. Care must be taken when examining the animals where poisoning is suspected because chemical contamination of the environment can occur, giving false results. Carcasses and unfixed tissues present particular dangers to personnel. They may harbour(1) infectious organisms and (2) cut or broken bones which cause injuries to human. The skins of animals also present dangers while handling museum specimens. Non infectious dangers arising from wildlife forensic necropsy are traumatic injuries from dead animals—Instruments, weapons and ammunition and snares, toxic poisons on or in dead animals including substances administered for veterinary purposes or as agents of euthanasia, or in proximity to them and ionizing radiation-- natural from certain soils, X- rays during radiological investigation of dead animals.

Forensic Necropsy

If a wild animal is found dead and has to be moved, take photographic record or drawing of its original position. This can be usefully supplemented by a chalk or paint outline of the carcass at the site (e.g. on a road or other hard surface) where it was located.

Advantages/ Disadvantages of removal carcass from crime scene.

Plan carefully and consult before removing the carcass. It may be necessary to station one team at the site to guard it, pending a decision. Local communities need to be consulted. Meteorological reports may be helpful.

Advantages	Disadvantages
Carcass can be protected from the loss or deterioration due to predators, scavengers (human) thieves, rain and other climatic factors as well as from poachers or other possibly dangerous, wild animals.	Since the scene is disturbed, the evidence may be destroyed or left in the field.
Enables a detailed examination to be performed under controlled equipment conditions by the best personnel and with the correct equipment.	Transportation may result in damage to bones and other material and jeopardize the chain of custody.
	Disturbance or separation of evidence. Additional contamination (DNA, bacteria, etc).
May improve the quality of certain tests where controlled conditions are necessary (to avoid contamination).Permits the involvement of elderly investigators who may not be able to carryout the work easily in the field.	A need for strict rules regarding packing and transfer of material.

Shipping Animal Carcasses

Note: Investigators should always contact the laboratory or facility involved before shipping an animal carcass from a wildlife crime scene.

Mark the bindle with case #, item #, date and your initials:

i. Carcasses that are suspected or known to be infected with disease.

ii. Carcasses which were used as pesticide baits.

iii. Significant number of carcasses (20 or more).

Carcasses should always be cooled down to preserve the tissues. Decomposition can hinder evaluation of the animal for diseases or evidence of trauma. Therefore rapid refrigeration and immediate shipping is essential. Any carcasses that are not significantly decomposed may be shipped to the laboratory without being frozen. Immediately after collection, place the carcass in a cooler to preserve the tissues. Triple bag the carcass and ship on blue ice within 24 hours of collection. Let the laboratory know that the carcass is fresh-never-frozen and call the laboratory to discuss shipment and receipt. Significantly decomposed carcasses and carcasses that cannot be shipped immediately should be frozen soon after collection in the field. If large carcasses are to be shipped frozen, they should be packaged and frozen in the container that is used to ship it so that the evidence assumes the shape of the container. Otherwise, it may be difficult to package.

Methods of cooling of Carcass

- Soak the fur in cold water with a small amount of detergent to aid in wetting of the skin, coat or plumage.

- Refrigerate the carcasses while awaiting examination.

- With very large mammals, perform the post-mortem as soon as possible since cooling of central organs will not occur sufficiently quickly to prevent autolysis. Making an opening in the abdomen may help lower the core temperature as quickly as possible.

- Place the carcass within a sealed plastic bag, clearly labeled, with excess air removed, and refrigerate if its body size allows.

- If sufficiently large refrigeration facilities are unavailable the carcass should be moved to as cool an area as is available.

- Refrigerate the carcass if post mortem examination is to be delayed until 72-96 hours after death.

Storage and Treatment of Wildlife Carcasses

i. If carcasses / tissues are chilled at +4°C , they can be preserved for few days, though autolysis continues to take place slowly but predictable.

ii. If frozen, carcasses/ tissues can be stored indefinitely, but freezing causes artifacts in terms of both gross and histological changes (Pathogens are generally not killed). Some allowance can be made for gross artifacts.

iii. If carcasses are fixed in formalin or similar fixative, tissues can be stored indefinitely (pathogens are usually killed) but affects appearance of organs – colour changes, Permits histopathology but makes DNA studies difficult. Careful interpretation is needed as in embalmed cadavers.

iv. Carcasses or tissues fixed in ethanol or methanol, can be stored indefinitely. There is little effect on DNA extraction. Though the appearance of organs (colour changes) is affected it permits histopathology but makes DNA studies difficult. Careful interpretation is needed as in embalmed cadavers.

In most cases a full necropsy is essential. Occasionally an incomplete examination is necessary either for legal reasons or because the rarity of the species means that the carcass is required for the research or display. Necropsy may not be complete for other reasons because in wildlife work portions of a carcass may be found or only certain derivatives are submitted for examination. Ascertain the background history, the species to be examined and its identity and provenance. Obtain adequate assistance. Plan attendance by other people Ex Police, SPCA inspectors who may need or wish to be present.

Forensic necropsy should be as thorough as possible. Failure to examine the brain of a mammal that suffered trauma to the head or to skin in a bird that possibly died of snake bite can lead to inaccurate diagnosis. When a need to preserve the body or parts of it, for research or display purposes, possibly on account of the rarity of the species arises, a cosmetic PME may be carried out (inflicting minimum external damage but obtaining essential information).An endoscopy may prove useful.

When there is large number of deaths, it is impossible to examine all the animals that are available for necropsy. Under such circumstances, sample of them are examined. Record how the sample selection was made to avoid criticism later.

Sometimes random samples or mixture of them are taken Ex. Severely affected and apparently normal specimens. Take larger sample size depending upon the time, practicality and the possibility that another expert may want to examine the material. Store some for a few days to be made available to other experts, if required or to be examined by the original pathologist later. Alternatively use more than one protocol so that some carcasses are subjected to full necropsy while others are not, only partly examined (paying attention to target organs and search for specific lesions). Record the main findings on a special group or flock form. This enables findings in different animals to be compared and contrasted rather more readily than if individual forms are used. The data can be hand written or prepared as computer spread sheet. It is essential to estimate how many animals have died or have been killed.

Differences Between Diagnostic and Forensic Necropsies

Diagnostic necropsy	Forensic necropsy
Determine the cause of death.	Determine the cause of death. Emphasis on comparing the history or suspicions from the law enforcement with findings.
Advice on preventing future losses.	Importance of external lesions / findings.
Emphasis on internal lesions	Preserve evidence
Importance of positive findings	Time commitment
Generally unchallenged	Chain of custody
	Photographic documentation
	Assume challenges to methodology, observations, objectivity and interpretations

Usually referred veterinary forensic cases

Unexpected sudden death—malicious poisoning, contaminated feed, death in clinic—minor surgery, boarding, grooming, ingestion of illegal substances, Shooting, environmental contamination and insurance cases, Starvation, animal abuse / cruelty, Veterinary malpractice (dispute between practices over case management), disagreements with drug, vaccine, feed companies, poaching or killing out of season, animal induced human injury and bestiality.

Steps to be followed in a Forensic Necropsy

- Collection of History including photos from crime scenes

- Identification (Case No., description, chain of custody)

- Radiographs (whole body when appropriate).

Forensic Necropsy Examination

Include comments on history, packaging on receipt of the carcass, carcass post mortem condition, scavenger damage, etc. Document these observations with Photographs. It is reasonable to review all ante-mortem information before initiation of the necropsy, including medical history (to identify pre-existing conditions that may be implicated in an animal's death). Witness statements of ante-mortem signs and clinical signs and Police reports of the events surrounding the suspected crime.

- Examine the external body surface (mouth, eyes, ears, skin, and anus) for lesions.

- Partially skin or skin carcass to note trauma, bullet or arrow wounds (to document trauma patterns) and predator or scavenging damage etc.

- Carry out the internal examination—Expose the thoracic and abdominal organs, examine in situ and document.

- Remove each organ and examine for lesions.

- Collect toxicology samples, bullets and other trace evidence.

- Document with notes, drawings and photographs.

- Submit appropriate samples for additional analysis.

- Document the collection of additional trace evidence and establish a "chain of custody" for each new piece of evidence.

- Compose reports describing findings and your opinion on the significance of those findings.

- Analysis and interpretation must consider the effects of PM conditions - pesticides degrade postmortem and bullets corrode the postmortem.

- At Necropsy table itself make sure that the photographs document the observations before moving on with necropsy.

- Anticipate as to how to demonstrate to a jury what was found and how it pertains to the case using photographs and/or x-rays.

- Record negative or normal observations in notes that would exclude alternative explanations by the defense which might be used to advance a different conclusion or throw doubt on the interpretation of the observations.

- Anticipate the opposite side's explanation and be prepared to counter or support it by the demonstration of facts.

- Establish what one can and can not say with the attorney who is presenting the results. Educate him concerning the science behind the observations or conclusions.

- Use professional quality photographs or displays to illustrate to the jury, the findings.

Precautions During Examination

- The table and environment should be clean (free from insecticides or disinfectants before the starting of examination).

- Allot separate laboratory Number to each carcass.

- Place only one carcass at one time. If more than one table is available or separation can be ensured (carcasses in separate trays) a production line system of examination may be considered.

- Cover the table with seamless sheet metal or with plastic or similar sheeting that can be changed between the examination (for easy cleaning).

- Clean and wash the table between each examination.

- All instruments and protective clothing should be changed between examination. If necessary several sets of instruments should be available.

- Minimize contamination during necropsy especially the spillage of fluid from one organ to another. Tying of loops of intestines or other organs with coloured string will help to reduce this risk and facilitate locating the organs.

Forensic Necropsy

- All samples for toxicological assay (blood, urine, tissues) should be placed in separate containers.

- The remains of each animal should be placed in separate storage bags at the end of PME.

- As in all forensic work, the side (not the lid) of each sample container should be marked with the relevant information. A useful extra precaution, especially when using alcoholic fixatives is to place a paper label with data inscribed in pencil, inside the specimen jar.

- People doing necropsy should not handle telephone. Use non-touch speaker, if telephone contact is needed.

Before PME commences, one must check whether there is

1. A requisition from the concerned to conduct postmortem.

2. Whether the person conducting the PME knows enough about the species to be confident to perform necropsy. If possible seek advice by telephone or e mail from a veterinary laboratory or experienced pathologist before deciding to embark upon any examination. If a more competent person is available in the near vicinity, consider submitting the carcass or some of them if there is more than one for proper investigation.

3. Whether one is clear as to what questions one will be asked, if the case reaches the court. If any doubts, delay the PME.

4. Confirm if the animal is dead. Some animals can go into a state of hypothermia and hypo-metabolism during hibernation. The ectothermic species—reptile, amphibians, fish and invertebrates, can appear dead because of low ambient temperatures and consequently a lowered metabolic rate. The traditional indicators of death are useful but not always totally reliable and even can be challenged in court. The ectothermic species that are "dead" may still have a beating heart which means they can bleed. Even in a mammal or bird sometimes blood comes out of haematoma or well vascularized wound, especially if the animal also has clotting defect. In ectothermic species, little information is available regarding the determination of death .These animals exhibit residual cardiac contraction even when clinically dead. In a carcass if the heart has stopped beating, it is considered as dead.

Important steps to be followed in the Field PME of a wild animal.

Obtain as much information as possible about the local circumstances on the death(s) and any relevant history. Take detailed notes. The background history consists mainly of information provided on the submission form, together with any additional data provided by the person or organization requesting the necropsy.

The general guidelines regarding the background information about a PM case is as follows. Always obtain some history before embarking on to the necropsy however brief, inadequate or unreliable may appear to be. Take note of history but do not rely on it. When there is an opportunity to take notes about the history do so. Do not depend on others who may have a less objective approach to the case. Important features of good history include: 1. Species of the animal (single or multiple), correct dates and times of incidents or findings. It can be very significant whether an animal was found sick or dead early in the morning or later in the day, 2. Food habits of the species involved, 3. Distribution of carcasses and death stance, 4. Postmortem condition of carcasses, 5. Evidence of scavengers and 6. Presence of potential baits, pest control activity and previous suspected poisoning events.

The appearance of animal carcass will be influenced by environmental factors that cause different types of damage. Ex.

1. Rough handling: Causes damage to periosteum, articular surfaces, epiphyses and the shaft of the bone itself. Similar effects are noticed due to spade, tractor or contact with hard objects, including other bones. Loss of hair and damage occurs to skin/hides and pelt with the same causes.

2. Exposure of water : Causes softening, leaching out of mineral, microbial and other attacks on the bones. Dampness causes growth of mould (fungi) often with secondary damage by invertebrates and vertebrates.

3. Contact with acidic or alkaline soil: Causes softening and pitting, reduction in weight of the bone.

4. Prolonged exposure to sunlight: Causes bleaching of bones and fading of colour of hair, skin and pelt.

5. Scavenging by vertebrate animals: Can cause or contribute to all the above effects on the bones, hair, skin and pelt.

Forensic Necropsy

6. Excessively dry conditions: Causes brittleness with fissure or breaks in the skin.

7. Primary attack by moths, beetles or other invertebrates: leads to destruction of hair and / or skin.

Following PME whole carcass or parts of them may need to be chilled, frozen or kept in fixative. As part of maintaining the chain of custody, these must be properly labeled. (Label with black letters on a white back ground or white letters on a blue o background). A label on a bag does not provide an identification of a specimen that is within that bag because the contents can easily be transposed. So apply a strip of tape to the limb or wrap adhesive bandage around the tail. Handle and store carefully the carcasses and all derivatives together with wrappings following PME. They may be examined later by prosecution or the defense. Important evidence may be from the animal's head or feet. Paper bags can be put over the extremities to preserve such evidence, plastic bags are not advisable because they gather condensation. The final disposal of carcass when the case is over depends upon the circumstances. It is wise to retain material for certain period in case it is needed. Return the body to its owner or keeper, to local community or for legal reasons to be passed to the appropriate agricultural or wildlife department. Retention of wildlife forensic material in museum and reference collections is important. Pending a court hearing, all wrappings that accompanied the carcass i.e. bags, boxes and padding should be retained. These are animal equivalent of "scene markers" and may provide important information about the circumstances of death. The key success in all forensic cases is 1. Thoroughness 2. use of standard procedure and 3. quality control and quality management. In necropsy work adherence to protocol, drawn up in the light of available information is essential.

Specimens for ancillary tests: Depends on the circumstances of particular case. Common ancillary tests are, histology, toxicological analysis, ballistics and DNA tests to identify the individual animals. Consult the laboratory in advance to determine the appropriate specimens, amount of specimens required and how it should be preserved. Avoid contamination especially for DNA and toxicological analyses. Use clean instruments for each specimen to avoid contamination and place each tissue into a separate container. For toxicological analyses, clean glass ware are used(acid rinsed).In cases of intoxication by unknown toxins, samples of stomach contents, intestinal contents, liver, kidney brain and fat should be collected in individual containers and held frozen. Label them carefully, seal and keep them in a manner that preserves the chain of custody. Because the rifled firearms leave characteristic markings on the projectile, bullets recovered from the carcass are useful for establishing the identity of the gun responsible for the injury. Shotgun pellets cannot be linked to a particular gun. Recovery of bullets and bullet fragments can be tedious especially in large animals and radiology or fluoroscopy are very useful. Once the bullet is established in a portion of tissue, careful dissection may be necessary. No tissue should be discarded until bullet has been found. Markings used for identifying the gun of origin are fragile and so care must be taken not to cut the bullet. Do not handle the bullet with metal forceps. Handle with gloved fingers or rubber covered forceps. These must be washed free of blood and tissues after recovery or these may harden and mask the markings. After washing put them in a padded container, label with case number, seal and initial them.

Trace Evidence

Each piece of trace evidence coming from a carcass which is collected for additional examination must be carefully labeled and tracked through the analysis process with a "chain of custody" (Sub samples from the carcass must have 'chain of evidence"). Seal all the evidences. Use certified laboratory with acceptable protocols for requested analysis and evidence handling. An analysts in a laboratory is subject to the same potential for court subpoena. Remind the laboratory that the material is from a legal case and that the analyst and any other technicians handling the evidence are subject to subpoena.

Consider the following carefully when submitting items to other agencies and individuals.

- Can the agency/individual undertake the work to a desired standard?

- Can continuity and security of evidence be maintained throughout the case?

- Do they understand the need for unique labeling of exhibits? For example, if samples are taken or separated from other exhibits

- If they intend to take photographs, have they been given appropriate advice on labeling, and how to handle film or digital images to ensure continuity of evidence.

- Are they aware of evidential issues such as unused material and the need to retain all notes, documents etc? Creating a file to hold all case information is a sensible system.

- Is the person undertaking the work prepared to attend court and would they make a suitable witness? Are they satisfied that they can respond appropriately to cross examination or defence evidence, which might question the validity of methods used and results obtained?

- Are any health and safety issues covered.

Collection of material for investigation

Carcasses of wildlife found in the field may be submitted for more than just cause of death and species identification analysis. Trace fibres found in the hair coat of mammals or the plumage may help point to individuals that came into contact with that animal before the agent arrived on the scene.

- In order to preserve fibres on the body, the animal should be carefully placed in the plastic bag before removal from the crime scene and should stay in that bag through the remainder of the packaging and shipping process.

- Solid items (such as traps and collars) found with the carcass may have fingerprint evidence on their surfaces. These items should be handled, packaged, and shipped with care.

- Pesticide-Laced Bait Carcasses: Talk with the laboratory about shipping pesticide-laced carcass or sampling the pesticide from the carcass. Some laboratories will not accept an entire pesticide-laced carcass.

- Oiled birds require special packaging in order to preserve the volatile chemicals within the feathers so that further testing can be performed.

- Oiled birds should be wrapped completely in aluminum foil.

- The foil-wrapped bird can then be placed in two more layers of plastic bags prior to shipping.

To prevent tearing of evidence bags leading to cross contamination of samples and possible injury to evidence handlers, use padding to cover beaks, talons, claws, teeth, broken bones, etc., of carcasses, especially where pesticide death or disease is suspected.

Additional trace evidence from a carcass: Bullets, bullet fragments and pellets, Burrs or other plant parts, Items from gastrointestinal tract, Hair samples, Tissue samples for chemical and histopathological analysis, Chemical residues, insects and DNA samples (saliva, blood stains etc).

- Issue separate identification Number based on original item. Take photograph or describe in situ if possible. Make sure analysis is done using 'court' acceptable techniques and evidence handling procedures. Give Reference to other reports when conclusions are based on another's analysis. Remember that the weakest link in the chain will be attacked by a good defense lawyer in order to discredit otherwise good forensic investigation. This could be the analyst in the laboratory to which the samples were sent for confirmation of your diagnosis. Remember that the other side also has a right to do their own analysis by their expert witness.

- Samples that are altered or destroyed in analysis (occurs in many chemical analysis) must be taken in duplicate to provide defense with sample for their own analysis. If you have not provided sufficient representative samples for them to work with, the case may be dismissed.

- Documentation of necropsy observations or the presence of trace evidence can be done with photographs and x-rays. Original digital photographs may be "locked" on a disc to assure that alteration can not take place. Write the report from the notes. Original table notes, drawings, etc. may be requested by the defense. Therefore make sure that there are no inconsistencies between your original documentation and your final report and/or testimony.

In case analytical procedures are not available locally, then, laboratory where the facilities are available are consulted. Before sending the specimens one should ascertain, whether the laboratory has experience in the legal and forensic analyses, understands the chain of custody and the scientist there is qualified to serve as an expert witness. The specimens are sent carefully in sealed containers along with request letter indicating the specific test to be performed to each of the specimens sent. The laboratory should be informed about sending of the parcel by registered post well in advance.

Special Techniques

Serologic and genetic techniques have a crucial role in wildlife forensic cases. In many cases, unrecognizable form of animal carcass or trace evidence (i.e., blood stains) is all that is available. It is necessary to identify meat samples stored in an individual's freezer or bloodstains in the bed of a truck. Antibody /antigen binding and electrophoresis techniques have been used for identifying animal species and to identify tissues in forensic cases. These methods are also used to help clarify the sequence of toxin exposure in cases of poisoned wildlife.

Shipping specimens: It is important to establish in advance that the laboratory chosen has experience with forensic or legal analysis, understands the chain of custody and the analytical scientist involved are suitably qualified to serve as witnesses. Do not send the specimens without making specific prior arrangements for the examination. Ensure that the specimens do not decompose in transit. Mark on the container as "Evidence". The pathologist should sign across the sealed tape. A covering letter and a request for analytical services required for each specimen is indicated. The sealed cover with letter is marked as document. Both are placed in a container on which the address is written and sent by parcel post or by courier.

Prompt and Timely Shipping

- Be sure to ship carcasses so that they arrive during work days.

- To control Odor Carcasses should be triple-wrapped in thick, heavy duty plastic bags and sealed tightly to prevent any leaks. Shipments can be stopped, inspected, and/or rejected by the shipping company if odors or leaks are detected.

- An odor control kit to take into the field for carcass collection may include the following:

- Plastic odor barrier bag Heavy-duty plastic bag, Odor-absorbing carbon cloth bag, Vinyl body bag, Carbon filter, roll of duct tape and zip ties.

Identification of the animal / carcass

Identification of an animal or the carcass of the animal is an important step in the vetero-legal (forensic) investigation. The question of identity arises in case of dead animal. When a dead body of the animal is examined, and based on certain data the conclusion is arrived at that it is the dead body of a particular animal. Identification of dead animal is established from the following data viz. breed, age, sex, colour of the coat, tattoo marks, and scars on the body of the animal.

Determination of Age

One of the parameters to establish the identity of the animal is by age. The age of the carcass is determined by close examination of dental formulae, bones, and horns. In case of females, the state of genitalia, scars on the ovaries, mammary glands and teats.

In case the identification marks are not available, then, measure the length of the animal from the point of shoulder to the point of hip, the height of the animal at the level of shoulder and the girth, behind the fore limbs. In case of horned animals, the length, colour (paint) of the horn and the distance between the horns can be noted. The colour of the coat of the animal can also be mentioned.

If no information on the age of the animal is available, then the same can be estimated approximately by observing the teeth in all domestic animals and the horns in case of horned cattle. In case of horned cattle the age can be estimated by observing the number of rings on the horns. The first ring appears on the horn at the age of 2 years. Subsequently one ring appears every year. Basing on this, the age of the cattle can be given by counting the number of rings and by adding 1 finally. If the age is estimated by this method mention should be made in the postmortem report.

Determination of Sex

At the time of inspection by regulatory authorities, the sex of the game carcasses is not grossly apparent.

Sex Organs

Identification of the sex organs obviously would be the easiest way to identify the sex of a carcass. The external organs usually are cut away in the field dressing process. Internally, the uterus, oviducts, or ovaries may be inadvertently left behind in the caudal abdomen. Sex is definitely female if the ovary is present and male if

Forensic Necropsy 169

testes are present. Sex is female if lactating udder , vagina, uterus , fallopian tubes and ovaries are present. If penis, prostrate and seminal vesicles are present the sex is male. A hermaphrodite or intersex is one in whom male and female characters exist in varying proportions. Hermaphrodites are of two types. (1) True hermaphroditism (both testes and ovaries are present) and (2) Pseudo-hermaphroditism which is of two types. Male pseudo-hermaphroditism (external genitals of female but have testes) and female pseudo-hermaphrodite (external genitals of male but have ovaries).

To determine the sex of a hermaphrodite, the following investigations are done.

A thorough external examination,
A thorough examination of internal organs,
Gonadal biopsy and
Study of sex chromatin.

Analyzing Trace Evidence

In some cases, it may be necessary to determine sex from blood or a butchered meat sample. In big game species, hormone analyses may be run on meat and blood samples to determine sex. These analyses are limited to animals that are in rut (males with elevated circulating levels of testosterone) and females with high circulating levels of progesterone (pregnant or having an active corpus luteum).

More recent genetic methods are available, such as PCR amplification of genes on sex chromosomes.

Condition of the body

*   Note signs of struggle.

*   Assess the condition of the animal and note the general condition of the fur.

*   Note any bite wounds or other signs of predation. If wounds are present, look for bruising and bleeding in the tissues near the wounds which would indicate that they occurred before the animal died. Otherwise these wound most likely were caused from the carcass being scavenged.

*   Look for broken bones, missing hair, broken or missing teeth or other signs of trauma.

- Look for and preserve any external parasites.

- Determine Nutritional Status of the Animal.

- Take weight (if possible) and/or body length and girth. Assess fat stores under the skin and in body cavities.

- Note the amount of fat around the heart and kidneys.

- Note the muscle mass of the animal.

- Note the amount of food in the digestive tract.

- Note the condition of the teeth.

- The condition of the carcass may be good or one may notice one of the following conditions.

- A carcass is said to be <u>poor</u> when the fat development is less with certain amount of fat in the orbits of the eye, and no fat in the sub-serous and intermuscular tissue (muscles are firm and darker in colour). The fat that is present appears normal and contains normal composition.

A carcass is said to be <u>emaciated</u> if there is no evidence of subcutaneous fat with all bones obviously protruding (abnormal retrogression in the condition) and with diminution in the size of its organs (liver, spleen and muscular tissue). There is alteration in the consistency of fat i.e. the fat has abnormal appearance-gelatinized (jelly like consistency and sickly yellow in colour). Since there is loss of inter-muscular fat, the muscles appear flabby and pale. If emaciation is coupled with anaemia, an increased amount of connective tissue may be noticed in the muscles leading to atrophy of muscular tissue. In young ones, the lymph nodes are swollen and edematous, the bone marrow is poor in fat and watery or replaced by slimy material. Gelatinization of fat is conspicuous in the fat over epicardial groove and at the base of heart. <u>Obesity:</u> is an abnormal and excessive accumulation of fat in the body stores (bulging fat all over with patchy pads around the tail). <u>Inanition</u> is a state of emaciation characterized by serous atrophy of adipose tissue, especially in the coronary grooves of the heart and the bone marrow and with some degree of atrophy of other body tissues, especially seen in skeletal musculature.

Forensic Necropsy

During the course of a forensic necropsy, the investigator attempts to determine

Time of death.
Cause of death.
Manner of death.
Mechanism of death.
It is imperative that environmental conditions are recorded whenever a dead body is found as well as the condition of the body.

Determining the Time of Death (TOD)

Estimation of Time since Death is done to determine whether suffering occurred in the period between assault and death. It follows that in such cases establishment of the time of death is important. With regard to farm livestock, the desire to establish the time of death may be related to questions on 1) duration of neglect, (2) failure to dispose of carcasses within statutory time limits, (3) uncertainties over death during transport, (4) suspected fraudulent insurance claims and (5) time of onset of serious husbandry problems (e.g. interruption of water supply to poultry houses). There are two basic approaches to estimation of time of death: (1) measurement of change that takes place at a known rate (e.g. rigor mortis, cooling of the body and putrefaction), and (2) comparison of the occurrence of events known to have taken place at a specific time with the time of death (e.g. extent of digestion of last meal).

Estimation of postmortem interval - The postmortem examination interval is the time that has elapsed between death and the discovery of a cadaver. The time of death is a frequent request by investigators. Establishing this can be helpful in eliminating certain people from an inquiry or may provide strong evidence that statutory requirements have been ignored. There is no single accurate marker of the time of death. The longer the period from time of death, the less accurate the estimates become (the body temperature, rigor mortis, levels of vitreous humour potassium levels, entomology studies-species and larval development and taphonomy i.e. Weathering of bones help in assessing the time of death). Body temperature may have some use in very recently killed animals and birds. Unfortunately, there are very few published data to help the veterinary pathologist establish the time of death in wildlife cases. The on set of changes that occur after death are influenced by stage of decomposition which is influenced by weather changes-environmental conditions-alternating exposure to sun and shade, wind exposure on a hillside, water immersion, dry heat, activity before death, body temperature before death, species and or size of the body, mechanism of transportation of animal, maggot infestation of the wounds, scavenger impact, amount of food in the stomach, and presence of systemic disease. The ability to accurately determine the time of death is inversely proportional to the length of time the post-mortem investigation is conducted. Many factors must be considered in estimating time of death.

Carcass examination can also be used to determine the time of death of the animal (although that can't be determined on trophy mounts and frozen carcasses). After an animal dies, its body undergoes a number of characteristic changes that can be used in estimating how long it has been dead. Time of death can be measured in a number of ways. The temperature of an animal approaches the ambient temperature after death. Charts and computer programs are available to estimate time of death using temperature data.

At best, the pathologist might estimate the time as falling into one of the following periods: less than 24 hours; several days; weeks; months or years. The state of vegetation around and under the body may also provide useful clues to the time the body has rested at one place. For example, whether the grass is stunted or discoloured under the body, or there is evidence of fungal growth. Such changes take days if not weeks to develop.

As in human forensic cases, determination of the time of death may be necessary to help place a defendant at the scene. In addition, wildlife forensic cases may involve the harvest of an animal during a non-sanctioned time of day, such as illegal night hunting. Many different techniques may be used and it is important to remember that there is no single accurate determinant of time of death. Of the following methods, body temperature loss and response to muscle stimulus are regarded as the most reliable.

Determining the time of death is an important aspect. In vetero-legal cases the autopsist should know the exact time of death because the court and the police may want to know. Though it is difficult, it can be determined by experience. This can be estimated by observing (i) body temperature, (ii) rigor mortis (iii) Electrical stimulus, (iv) Changes in the eye (v) Hypostatic congestion or cadaveric lividity and (vi) Signs of decomposition / putrefaction.

Methods for Estimation of the Postmortem Interval

1. Temperature based methods 2. Postmortem chemistry, 3. Electrical stimulation of muscle and nerve, 4. Gross appearance of body (Rigor mortis, Eye shape, colour and luminosity etc and Decomposition), 5. Histopathology and electron microscopy, 6. Radiology, 7. RNA and DNA analyses, 8. Entomology and 9. Environmental and associated evidence.

Wildlife crime investigations are less concerned with issues of suffering because they are usually focused on regulatory matters (e.g. 'out of season' shooting of game animals, poaching and breaches of statutory time limits). Many countries have enacted laws to provide protection of wildlife through limiting hunting to specific periods of the year and by requiring trappers to visit their traps and snares at least once in any 12 or 24 h period. Estimation of the post- mortem interval can be helpful during investigation of breaches of these regulations. Establishment of the approximate time of death of protected species (e.g. badgers) can serve the same purpose of determining alibi and opportunity, just as for human and companion animal deaths.

While determining the time of death, one should keep in mind (1) the atmospheric temperature and (2) season. These two influence the rigor mortis and putrefaction

i. Body Temperature

It may be a useful indicator of the post-mortem interval during the first 12 to 24 hours following death. In mammalian species, intranasal temperature (measured at caudal extent of the nasal cavity) and the temperature at the center of muscle mass in the hind leg usually are taken. In smaller mammals, rectal temperature is used (3 inch insertion).

For avian species, measurements usually are taken by intra-thoracic or cloacal insertion. 3 inch insertion for both).

Loss of body temperature is dependent on the following variables.

Initial body temperature - The initial body temperature is fairly constant for a given species. However, body temperature may be elevated by pyrexia (disease) or hyperthermia (physical exertion).

Ambient temperature and humidity - Ambient temperature and humidity greatly affect the cooling of a carcass. These parameters are relatively easy to measure. It is necessary to account for weather fluctuations over the relevant time period. Water temperature may be a consideration, in the case of waterfowl.

Body surface area - Body surface area is relatively constant for a given species.
Body mass - Investigators may estimate body mass by measuring the girth of the thorax at the level of the heart (heart girth).

Carcass handling - The manner in which a carcass is handled during the post-mortem interval also will influence body temperature. If a carcass is field dressed, the body mass is greatly decreased by removal of the internal organs and the surface area is increased. These changes will increase the rate of loss of body heat. Skinning a carcass will produce a similar effect. If the carcass is transported in a vehicle, air flow will cool it much more quickly than exposure to still air. Typically, intranasal temperature drops faster than intramuscular temperature during transport. Finally, carcasses may be insulated, either by being piled with other carcasses or being kept at a higher ambient temperature. Taking several temperatures over several hours may improve accuracy of the time of death estimate.

The rate of cooling of the body is referred to as álgor mortis" It is a single indicator of the time of death during the 1st 24 hours post-mortem. The intra-abdominal temperature may be measured either per rectum or the intra-hepatic/ sub hepatic temperature through abdominal stab (which allows insertion of the thermometer on to the under-surface of the liver). Readings may be taken rectally or in the liver. Make a small cut in the skin and insert the thermometer into the liver or under a lobe to avoid damage. A chemical thermometer 25-30 Cm long with a range from 0-50° Celsius is ideal. When taking the temperature a special thermometer is needed to register low temperatures. There is an initial temperature plateau that occurs in the first 30 minutes to 5 hours. The normal rate of cooling is 1.5 degrees Fahrenheit loss/hr (rectal) at 75 degree environmental temperature. Take two reading in an hour to get the rate of cooling. After death, the body cools unless the ambient temperature is greater than the body's temperature. Algor mortis is affected by several factors, including ambient temperature, ante-mortem activity levels, body insulation (hair/adipose tissue), immersion in water, infliction of significant tissue trauma/destruction, local movement of air, or underground burial. It can be affected by temperature prior to death, size of the body, dehydration, obesity, edema, body

Forensic Necropsy

position (curled vs. recumbent), hair coat, humidity, and wind, cover. Air movements accelerates cooling of body. A cadaver cools more rapidly when immersed in water than in the air because water is a better conductor of heat than the air. Similarly the body cools rapidly in humid atmosphere.

Time elapsed since death can be calculated from the formula
Normal animal temperature –Temp. at the examination time = Rate of cooling (hours)

ii. Rigor Mortis

Immediately after death, the voluntary and involuntary muscles relax totally (known as primary muscular flaccidity in forensic medicine), due to loss of tone. This leads to drop of jaw, dilatation of eye pupils and the heart.

Rigor mortis is the progressive stiffness of muscle groups that occurs following death due to the depletion of ATP and phosphocreatine. In general, full body rigor takes about 6 - 12 hours to develop and is lost by 24-48 hours (Appendix 4). Relaxation of rigor occurs by autolysis. The extent of rigor is measured by flexion of joints. The sequence of detectable rigor has been experimentally determined for several species. In cervids, rigor begins with the jaw followed in order by the knee, elbow, tarsus, neck, and ends with the carpus. Investigators typically test both sides and use the side that is most advanced for their estimate.

Disappearance of rigor mortis depends upon the rapidity with which decomposition occurs. The appearance of rigor mortis and putrefaction are hastened by external temperature.

1. In warm air and moist climate the rigor mortis is rapid and of shorter duration.

2. In animals died with violent exercise (racing, fighting, and struggling) and violent muscular contraction (tetanus, strychnine poisoning) before death, the rigor mortis is hastened and of shorter duration. In these putrefaction is also hastened.

3. In emaciated carcass it is weak and appears early and of shorter duration.

4. In low temperature (winter- cold and dry climate) rigor mortis is retarded.

5. In well-nourished and healthy carcasses it is well defined and lasts longer and appears late.

6. In bodies immersed in water the rigor mortis commences rapidly and disappears late.

The presence and extent of rigor mortis can be detected by manipulation of selected joints. The timing of the above observations is affected by several variables, including ambient weather conditions, size and surface area of the carcass, handling of the carcass (intact or field dressed, method of transportation, etc.), and behavioral state of the animal at the time of death (relaxed or under stress).

Limitations

First, the assessment itself is relatively subjective. Many variables may affect the rate for rigor. The process occurs more slowly at cooler temperatures and is more rapid in animals that died following physical exertion. Rough handling of a carcass may delay or prevent the onset of rigor mortis. It is recommended to avoid the muscles adjacent to wounds and legs that are stretched when a carcass is hung. Also, freezing must be differentiated from rigor in some situations.

Time of Death (TOD) Estimates in Animals

Not stiff, Warm	0-3 hours
Stiff, Warm	3-8 hours
Stiff, Cold	8-36 hours
Not stiff, Cold,	>36 hours

iii. Electrical Stimulus

The ability of muscle groups to respond to electrical stimulation decreases with increasing time after death.

Electrical stimulation provides a less subjective method for measuring the availability of ATP; however, this technique is affected by the same processes as rigor mortis. In addition, severe brain injury may affect the response. The apparatus for performing the test is a simple circuit powered by the ignition system of a vehicle. The response of various muscle groups are ranked as very good, good, fair, and poor. Responses usually remain good for 4 hours or less following death.

Muscle groups stimulated in the deer include the eye, muzzle, ear, proximal forelimb, epaxial muscles, tail, exposed inner thigh, tongue and exposed flank. This technique may be applied to waterfowl as well. If a reaction is detectable in the wing, it usually indicates that a duck has been dead for less than 1 hour. The same finding suggests a goose has been dead less than 2 hours. Muscles tested in water fowl include extra-ocular, bill, epaxial, tail and wing.

iv. Changes in the Eye

Changes in the appearance of the eyes can also be used.
Examination of the eyes: Soon after the death, the eye loses its luster. The corneal reflex is lost. The cornea becomes opaque. In delayed death the eyes are closed. Such a condition may be present before death in uraemia. The cornea may retain transparency for some time after death in hydrocyanic acid and carbon monoxide poisoning.

The pupil is roughly oval in cows and horses. When it contracts it becomes elliptical. The iris is commonly light or yellowish brown in colour. Rarely it has a blue tinge. It is round in shape. In pigs the iris is dark-greenish brown or yellowish brown and is oval in shape but becomes circular when dilated.

Ophthalmic examination also may provide an indication of the post-mortem interval. In cervids, parameters that are examined include pupil diameter (vertical), tapetal luminescence, and intraocular fluid character. The general sequence of changes that have been reported for the cervid eye is given in the Appendix 5

v. Hypostasis or cadaveric lividity or postmortem staining

After death , there is gravitational pooling of blood in the veins and capillary beds of the most dependent parts of the animal body, following cessation of the circulation leading to a dark purple discoloration of the skin i.e. the discolouration of the skin due to accumulation of fluid blood in the most dependent parts. This is called post-mortem Lividity / Hypostasis / or livor mortis or postmortem suggillations (Appendix 3). This appears as irregular livid patches in the most dependent parts of the body in the subcutaneous tissues on the side on which the animal has been lying i.e. back, lumbar region, posterior aspect of the limbs and shoulder. It is well marked in asphyxia. Blood does not clot after death from asphyxia or when large quantities of saline are infused. The blood clots readily in deaths from some acute infectious diseases. It moves if the body is moved before 4 hrs. It is patchy, well marked by 4 hours and gets fixed in 6 hrs. It differs from contusion, which is elevated, occurs anywhere and does not move if the body is moved. The postmortem staining will be marked in asphyxia and in certain infectious diseases. By knowing the distribution of hypostasis, we can know whether the body has been disturbed or not.

The hypostatic congestion is a special form of hyperaemia that occurs in the dependent parts of recumbent animals especially in the lungs. This should be differentiated from postmortem congestion, which occurs in the lower most parts of the body under the influence of gravity. The colour of hypostasis indicates the cause of death (purple- Asphyxia, Cherry red-CO, HCN poisoning, Chocolate colour- Nitrate/ nitrite poisoning).

vi. Putrefaction and decomposition

Accurate differentiation of *ante-mortem* and *post-mortem* changes is important in all necropsies but perhaps never more so than in a forensic case.

Progression of post-mortem changes depend upon such factors as

- Species of animal.

- Health status before death.

- Presence or absence of ingesta in the gastrointestinal (GI) tract.

- Manner of death.

- Body temperature at time of death.

- Environmental temperature and humidity.

- Location and position of the body.

- Handling and storage following retrieval of the body.

Useful clues to the animal's history, time since death and/or the environmental conditions to which the carcass has been exposed may be provided by

- Presence or absence of *rigor mortis*, *livor mortis* (hypostasis, pooling of blood, etc.) and *algor mortis* (cooling).

Forensic Necropsy

- Appearance of organs and tissues.

- Evidence of predation/scavenging by other animals such as dogs: this may be *ante mortem* or *post mortem*.

- Infestation by maggots, carrion, beetles, etc.

Organs and tissues undergo *post-mortem* autolysis at different rates (starting with those organs that rapidly show breakdown and finishing with those that show relatively little decomposition).

- Retina

- Brain

- Testis

- Gastro-intestinal (GI) tract

- Pancreas

- Liver

- Kidney

- Skin

- Skeletal and heart muscle

- Uterus

There are also species differences. For example, the skin of fish and amphibians tends to autolyze sooner than does that of a mammal or bird. Although, as a general rule, autolysis is more rapid at a higher temperature, caution must be exercised.

This begins from 6 to 36 hrs depending upon the factors mentioned below. It appears as greenish discolouration of abdominal wall within 24 hours or 2 or 3 days depending upon the season of the year. Putrefaction follows disappearance of rigor mortis but not always. The body begins to emit foul odour owing to gradual formation of gases of decomposition (Hydrogen sulphide, methane, carbon dioxide, ammonia, hydrogen phosphate etc). After 18-36 hours after death, gases collect in tissues, cavities and hollow viscera under considerable pressure with the result the body becomes bloated and distorted, eyes are forced out of sockets, tongue protrudes and lips swollen and everted. Frothy fluid or mucus is forced out of the mouth and nostrils. The abdomen is distended. On opening the gases escape with loud noise. Discolouration appears early in cases of suppurative peritonitis. The characteristic features of putrefaction are, colour changes and development of foul smelling gas.

Factors

1. Condition of the animal at the time of death. Animals with fat putrefy quickly.

2. Temperature and moisture of the atmosphere.

3. Species of the animal. Sheep putrefy quickly than short-coated animals.

4. The manner of death and state of the body. Animals, which die of abdominal diseases, putrefy quickly. But it is delayed in those dying of arsenic, antimony, strychnine and chloroform poisoning. It is also delayed in carcasses buried in dry soil and submerged in water. The appearance of rigor mortis and putrefaction are hastened by external temperature, violent exercise (racing, fighting, struggling) and violent muscular contraction (tetanus, strychnine poisoning). It is retarded in animals with emaciation and anemia.

Even from the putrefied carcasses approximate time of death can be estimated by the presence of eggs or pupae or adult flies. Flies are attracted to the body and they lay eggs on the body especially on the wounds and natural orifices. The eggs hatch into maggots or larvae in about 24 hours, the maggots become pupae in about four to five days and the pupae become adult flies in about 3 to 5 days. So from its stage of development a rough estimate of the time of death can be made (forensic entomology).

Postmortem autolysis

In autolysis of the carcasses, muscles appear pale, soft and watery. The epithelium of the rumen, reticulum and omasum show desquamation. Putrefaction sets in early in animals died due to electrocution and lightning. Most carcasses will have some autolysis, but diagnostic tests can still be performed if tissues are properly handled. (Handle autolyzed tissues for histopathology very gently).

Forensic Necropsy

- Hold tissues at the edges only. Cut with a sharp knife or scalpel. Quickly place in formalin).

- Freeze or refrigerate samples as soon as possible for infectious disease or toxicology testing.

Autolysis can cause many artifacts in tissues that can be confused with a disease process. However, it is always best to take a sample from an area that looks abnormal rather than assume that the change was caused by autolysis. Histopathology will be able to distinguish between true lesions and post-mortem changes.

vii. Forensic Entomology

Entomology can be very useful, as the degree of decomposition and any maggot or beetle infestation may be important in establishing an approximate time of death. By understanding the time and sequence of various stages as bacteria and insects attack the body, the entomologist can often make an accurate estimate of the time since death. In addition, many different species of insect attack a corpse at different stages of decay and even prefer different tissue types. Since some insect species are only found in certain geographical regions, climates or locations this may indicate that a body has been moved from another location. These items should be retained for examination by an appropriate entomologist. Advice should be taken at the earliest opportunity on how the items should be handled and transported.

Insect analysis is also a good method for determining TOD. Maggots can aid in determining location of death and provide DNA and toxicology evidence. Maggots can help determine the time of death by providing the post mortem interval. Flies lay eggs during certain environmental conditions, at certain times of day after an animal has died depending on the species of fly. These eggs then hatch into maggots based on environmental conditions. The larvae develop at a certain rate, depending on the species and environmental conditions, and can be aged by a forensic entomologist. Blow flies are attracted to the body postmortem so by dating the time of colonization (laying of eggs), the time of death can be estimated. It is important to note that in some cases, maggots maybe found on live animals, known as myasis. This is usually due to fecal soiling or wound necrosis present on the animal that attracted the flies. In this case, the time estimate will be for the time of trauma. Other insects are forensically important such as beetles which feed at different times postmortem.

A sample of all insects, pupae and pupa casings on the body should be collected noting the location on the body they were found. Forensic entomology measures postmortem interval by assessing the life stages of necrophagous insects. This field has not been applied extensively in veterinary forensics (Appendix8).

Determining the Time of Death

This cannot be directly determined at the crime scene simply by taking the liver temperature or feeling the stiffness of the body. Time of death (TOD) is calculated by 1st estimating the postmortem interval (PMI) or the amount of time that has elapsed since death. Determining the TOD and PMI can be crucial to a successful prosecution. Determining the TOD can help prove that the suspect's alibi is false or can support a witness' statement. PMI is estimated by integrating information from a variety of changes that the body undergoes after death, including rigor mortis, algor mortis and decomposition. Decomposition of a body is accompanied by the colonization of that body by certain insects. These insects colonize a dead body at specific times in the process of decomposition, in specific patterns and they grow and develop at specific rates, depending upon the climate. If the investigator documents entomological activity and climate conditions and properly collects samples, a forensic entomologist can estimate the TOD of the victim. Entomological estimation of TOD is the most accurate way to determine the PMI if the body has been dead for longer than 3 days. In most cases the PMI is calculated using each of these individual techniques. These numbers are then compared to ensure the most accurate estimation of TOD. Each of these techniques will not elicit one time of death, but a range of possible lengths of PMI. The overlap between ranges that occurs when the techniques are compared helps to narrow down the length of PMI and approximate TOD.

Cause of Death

The disease, injury, or abnormality that alone or in combination is responsible for initiating the sequence of functional disturbances that ends in death. Examples include gunshot and blunt trauma from vehicle collision. Example: 1. A wild animal is struck by an automobile, thus fracturing its mandible. The animal eventually dies from starvation. The cause of death is blunt trauma due to vehicle collision because this event precipitated all other processes.2. Ex: Gunshot wound to the wing of an eagle which eventually dies in an emaciated state due to Aspergillosis.

Forensic Necropsy

Manner of Death

The manner of death is the way in which the cause of death came about with special reference to social relationship and personal causation. Ex. Death of eagle due to pesticide poisoning after eating legally poisoned rodents (accidental) or eagle dying due to pesticide poisoning after eating on carcasses of sheep poisoned with pesticides.

A classification of the way in which the cause of death came about with special reference to social relationship and personal causation. 1. Example: Accidental vs. Illegal death. A white-tailed deer that is hit by a motorist out of season is an example of an accidental killing. A white-tailed deer that is shot out of season for trophy purposes or meat is an illegal killing. 2. Ex.1. Eagle dies of pesticide poisoning after eating legally poisoned rodents (accidental). Ex 2. Eagle dies of carbofuran poisoning after eating on carcasses of sheep laced with carbofuran (intentional).

Mechanism of Death

The mechanism of death is defined as structural or functional change that makes independent life no longer possible (myocardial laceration from bullet wound results in internal bleeding and finally tissue anoxia). 2. Severance of the cervical spinal cord by a bullet resulting in respiratory paralysis.

Modes of Death

In all forms of death the fundamental pathological alterations are uniform but the degree of change varies. In sudden primary cardiac failure, the visceral congestion is less or absent and in deaths occurring slowly, the congestion of viscera is more. Sudden or unexpected death of an animal may occur due to unnatural causes such as violence or poisoning or even from natural causes. All cases of unnatural deaths have to be investigated. Under certain circumstances natural deaths (which occur suddenly in apparently healthy animals under suspicious circumstances) form the basis of vetero-legal investigation. In such situations, the possibility of death as a result of disease and injury together has to be kept in mind. In such cases the veterinarian should not certify the cause of death without conducting the postmortem examination even where there is a strong evidence of death due to disease. The various causes, which produce sudden death, are the diseases of cardiovascular system (acute myocarditis, fatty degeneration of heart, myocardial infarction , rupture of aneurysm, aortic valvular disease of left ventricle (left ventricular failure), right ventricular failure associated with chronic emphysema and other diseases of lungs, diseases of pericardium, congenital anomalies etc).

Death is classified into somatic or systemic death (complete loss of sensibility and ability to move and complete stoppage of functions of brain, heart and lungs "Tripod of Life") and molecular death (death of tissues and cells individually, which takes place after some time after stoppage of vital functions). All kinds of death (either natural or accidental) are due to three primary reasons. These are coma, syncope and asphyxia (Appendix 6).

Not all *post-mortem* examinations provide a diagnosis, even in standard medical or veterinary work.

Examination of the Animal

A. External Examination

Lividity is not often seen in the external skin of dogs and cats. It is more likely to be seen on the buccal mucosa, internal organs and body walls. This can help determine the position of the body after death. Contact points with the ground or other objects will exert pressure on the body and prevent blood from settling at these points, resulting in blanching that may provide information about the contact surface.

Weigh and measure the animal. The animal should be weighed prior to necropsy and the weight recorded. If the carcass is incomplete,(often whole carcass may not be presented for necropsy) the portions that are present should be described and the absence of remainder should be noted. The limbs should be manipulated to assess the stage of rigor mortis and the presence of fractures or dislocations.

During the external examination, signs of traumatic injury (i.e., puncture wounds, lacerations), nutritional condition (emaciation versus obesity), and natural disease are recorded. An anatomic body chart is helpful to annotate significant information. In some cases, a fine tooth comb can help collect evidence that may be loosely attached to the body, such as botanical, entomological, or trace evidence, which may provide valuable information about the case. Collection and packaging of evidence should follow recommended guidelines based on the evidence type. Generally, full body radiographs should be taken and examined prior to necropsy to evaluate abnormalities of the skeletal system, identify projectiles, and locate any microchips. Radiographic interpretation often helps identify sites on the skeleton where gross examination of specific bones is warranted to identify the nature of injury or traumatic fractures. Each radiograph should include the case number and animal's identification information, labeled with correct left and right markers. Failure to adequately identify and label radiographs may result in their inadmissibility in a court of law.

Examine the external orifices, identify the parasites (as they may have come from another species, and this can provide useful forensic information. Ex fleas found in foxes came from the prey), lesions or abnormalities. Retain samples, fixed or fresh as necessary. Check important structures (preen gland in birds, teeth in rodents). Photograph or draw as much as possible.

External examination in human forensic autopsies is of great value because the injuries are readily visible on the skin of the individual. But in the animal necropsies the superficial features are masked by the plumage or pelage. Penetrating wounds (gun shot wounds) caused by small calibre rifle bullets or shotgun pellets are difficult to find externally. The pathologist should record any evidence of skin abnormalities, or bleeding from the body orifices or soiling with blood, excreta, exudate or foreign material. Samples of foreign material such as oil on the plumage or pelage should be collected in suitable containers and preserved for possible analysis. In large birds, the entire body should be examined carefully for burned or singed feathers or skin as a result of electrocution. Remove the hair, by clipping and feathers by plucking to detect superficial lesions. Skinning is important in forensic necropsy to detect injury to the integument and superficial tissues. Most pathologist remove the skin after internal examination but some do it at the outset. Carefully skin the entire animal because the traumatic injuries due to projectile and bite wounds, trap or snare marks are readily visible on the flesh side of the carcass. Skinning is done to detect (1). Edema or haemorrhage, indicative of traumatic injury, (2). To detect wounds due to trap and snares in the deeper tissues than on the furred, feathered or scaled surfaces. Snare injuries in mammals are usually common on the limbs , (3). To examine for subtle marks caused by leg-hold traps in furbearers and large raptors in the distal limb. If trap injury is suspected, the leg should be radio graphed to see the fracture in the small foot bones, (4) In all cases of suspected gun shot wounds because bullets that have lost energy in passing through the body are unable to exit through the tough elastic hide. This is usually indicated by the presence of subcutaneous haemorrhage without skin perforation on the side of the body opposite a punctured wound. The bullet in such cases are lodged immediately under the skin and lost easily while skinning. If traumatic injuries are suspected, intact carcass of birds and smaller mammals and affected portions of the carcass of large animals should be radio-graphed. Correct identification must appear in the radiograph and (5). Skinning permits the assessment of s/c fat and the appearance of the musculature.

Then the standard necropsy procedure is followed viz. systematic examination of positional relationship among organs and of the size, colour, shape and texture of all organs including the brain and spinal cord. All the hollow organs are opened to inspect the lumen and contents. The parenchymatous organs are incised. The process of opening and examining body cavities and organs alters their relationship and appearance so that the observations should be made and recorded as the process proceeds. The abnormalities should be photographed as soon as possible since the appearance of tissue changes and lesions may be destroyed during the necropsy procedure Ex. If the interior of the thoracic cavity is not examined immediately after opening it may be impossible to determine if the blood was present initially or seeped in from other tissues during necropsy. It is essential that all abnormalities, no matter how trivial they may appear, are recorded. Record negative, as well as positive, findings. Abnormalities or lesions should be described in absolute units (centimeters, grams) rather than in relative terms (enlarged, smaller than normal). The location of injuries and abnormalities should be measured and recorded in relation to readily identified anatomic landmarks. Outline drawings prepared in advance or at the time of necropsy, are particularly useful for recording the location of lesions such as traumatic wounds. These drawings become part of the permanent record of the case.

All organ systems should be examined and all abnormalities described, even those usually regarded as incidental in a standard necropsy. Finally, normal findings should be described as well. These extra steps provide written support that a complete and thorough necropsy was performed.

B. Internal Examination

During the course of the necropsy, it is important for the pathologist to have the foresight to collect the appropriate samples in the correct manner. In addition, the

Forensic Necropsy 187

samples would have to be stored in the proper container (acid-rinsed glassware in this case) and shipped in a secure manner to the appropriate laboratory. Another important consideration in forensic necropsy is the use of experts in other disciplines to perform necessary analyses (Example-forensic firearm specialist would assess projectile character and make conclusions about ballistics)

As soon as the abdominal cavity is opened, examine the organs before removal for anomaly / malformation, Rupture, Volvulus, intussusception, twist, Hernia and displacements

Note the position of the organ in relation to the other organs.

- Record any lesions or abnormalities and note whether alimentary track contains food or other material.

- Check important structures. Retain samples-fresh or fixed, as necessary including frozen-lungs, liver, and kidney for toxicological investigations.

- Photograph or sketch as much as possible.

- PM scavenging has to be distinguished from ante-mortem predation. Carcass of a wild animal is very likely to be scavenged by other animals. Some cases of scavenging can be spectacular.

- Examine Pleural, pericardial and peritoneal cavities Note the colour, consistency and the quantity of fluid present.

In the organs note

1. Colour

2. Size and shape and

3. Consistency.

The remainder of necropsy consists of systematic examination of the positional relationship among organs and of the size, colour, shape and texture of all organs including the brain and spinal cord. All organs should be opened to inspect the lumen and their contents. Incise parenchymatous organs.

The necropsy report should be written immediately after the completion of necropsy from the rough notes or dictation made while doing the necropsy. The original notes should be retained by the pathologist until the court proceedings are completed. In case some ancillary tests are to be done, a preliminary necropsy report is prepared mentioning the list of samples retained for analysis along with the list of photographs and radiographs taken. A final necropsy report is prepared when the results of all the tests are available. This should contain the observations of the pathologists, results of ancillary tests, list of diagnoses, list of specimens taken and their disposition as well as the pathologist's interpretation of the meaning and the significance of the findings. In case there are some inconsistencies or if some results are equivocal this should be pointed out and an alternative explanation should be discussed. After preparing the final report the pathologist should read the report carefully and sign every page before it is sent. The report is confidential and should be given to the authority who requested the necropsy examination. The pathologist should keep in his possession all the documents pertaining the necropsy of each case which are not returned to the enforcement authority including the necropsy report, photographs, results of ancillary tests and physical specimens. These should be kept in his custody in a secure manner until the case goes to the court or until the enforcement authority informs that the case has been closed and that the information is no longer required.

History, age, nutritional state, concurrent disease or injury, etc. should be documented. These findings may be especially relevant in cases where the cause of death is debatable, such as in toxicoses.

Data given on the evidence tag should be noted in the report including the name of the submitting officer and agency.

All descriptions should be concise using a non-technical terminology. All conclusions should be supported by relevant information and alternatives discussed when results are equivocal.

It is important to remember that the necropsy report ultimately may have to be understood by non-medical people.

The most commonly encountered forensic cases requiring pathological evaluation are Gunshot wounds, trauma wounds and acute poisonings. The pathologist should recognize and interpret these conditions in the forensic context and be able to document and collect appropriate supporting evidence.

In wild animal forensic necropsies, traumatic wounds due to gunshot or predator wounds form very important part. Pathologists who perform necropsies in such cases should understand the basic ballistics and should be familiar with the type of injuries produced by different firearms as well as those caused by arrows, the predators in the areas and various forms of blunt trauma. In all cases of gunshot wounds it is always necessary to trace the path of the projectile through the body since it gives the relative position of the animal and the shooter. The bullets usually pass relatively in a straight line through tissue but the impact with the bone may change the direction dramatically. If the bullet fragments , there may be several diverging paths. The radiographs in such cases (gunshot wounds) are invaluable. Small fragment of metal may mark the path of the bullet in tissues and the nature of the projectile is clearly visible in the radiograph. The entrance wounds are often smaller in diameter than the exit wounds due to distortion of bullet as it passes through the tissues. If the bullet fragments, the individual exit wounds may be very small. The diameter of the wounds in the skin and soft tissue is not the reliable indicator of the size (calibre) of the causative bullet. When the bullet passes through any site where flat bones are present like skull, pelvic bones or scapula, special attention has to be paid because the resulting fracture may provide the evidence of direction of travel of the projectile. Many wild animals have bullets or shotgun pellets embedded in their tissues as a result of earlier unrelated events. It is important to relate the location of bullets in the wild animals to the presence of appropriate recent wounds.

Sometimes it may be necessary to establish the relationship between the separated portions of the tissue (ex. Recovery of head of the game animal at one place, the viscera at the 2nd place and the carcass at the 3rd place) belong to a single animal. This can be accomplished by careful matching of the structures by using radiography and dissection of the parts and features such as apposing cuts and fractures on the two ends of the bones. Bones have to be cleaned of flesh to demonstrate fractures or cut marks. If bones are cleaned of flesh by boiling, individual portions must be carefully wrapped in cheese cloth and labeled to avoid intermixing of specimens during boiling process. Direct comparison of DNA extracted from the individual portions can also be used.

Investigation at the scene of crime/scene of occurrence in cases of poaching or unnatural death of wild animals -

The prime objectives of crime scene investigation in the incidences of poaching or unnatural death of wild animals are to establish (i) how the animal was killed, (ii) where the animal was killed and (iii) what could be the possible date and time of killing, and (iv) who killed the animal. Demarcate a reasonable extent of the area surrounding the place where the carcass or the body parts are lying for conducting search to trace evidence. The whole area thus demarcated is to be construed as the scene of crime. It should be kept in mind that anything lying within the scene of crime could be valuable evidence depending on the case. Therefore, no material or object lying within the scene of crime should go unnoticed and unrecorded.

In cases, where a carcass is found

Post-mortem examination of the carcass should be done by a veterinary surgeon or by a team of veterinary surgeons. The Post-mortem report submitted by the veterinary surgeon or a team showing the cause of death, possible time of death etc., should be sent to the court along with the complaint.

In cases where fire arms are used

Retrieve the pellets or the bullets from the carcass during post mortem examination to identify the type of the fire arms used. Pellets or bullets should be sent to the forensic science laboratory for examination by the ballistic expert to ascertain the type of fire arm used. Since forest officers are not empowered to investigate offences under Indian Arms Act, therefore, whenever weapons are recovered or suspected to be used, intimation to this effect should be given to nearest police station.

In cases where If only body parts like bones, pieces of flesh, hairs blood etc are found

The same should be collected, sealed and preserved preferably by the veterinary surgeon following all the legal formalities. The body parts thus sealed and preserved should be properly labeled and signed by the veterinary surgeon(s), independent witnesses, forensic expert and the Investigating Officer. The body parts or samples of the articles thus collected should be sent to the Wildlife Institute of India (WII) or other such expert institute to identify the animal killed.

A list of Offices of Wildlife Crime Control Bureau

BHQ: All India Jurisdiction

Additional Director, Wildlife Crime Control Bureau, 2nd Floor, Trikoot - 1, Bhikaji Cama Place, New Delhi -110066.

Southern Region

- **Andhra Pradesh, Karnataka, Tamil Nadu, Kerala, Puducherry and Andaman & Nicobar Islands**

Regional Deputy Director Wildlife Crime Control Bureau / Southern Region, C2A, Rajaji Bhavan, Besant Nagar, Chennai-600090. Eastern Region

- **Assam, Bihar, Meghalaya, Manipur, Mizoram, Nagaland, Sikkim, Tripura, Arunachal Pradesh, West Bengal**

Regional Deputy Director, Wildlife Crime Control Bureau / Eastern Region, Nizam Palace, 2nd MSO Building, 6th Floor, A.J.C Bose Road, Kolkata-700020. Western Region

- **Goa, Gujarat, Maharashtra, Daman & Diu, Dadar & Nagar Haveli**

Regional Deputy Director Wildlife Crime Control Bureau/ Western Region, Room No.501/B, V th Floor, Kendriya Sadan Building, CBD Building, Belapur, Mumbai-4000614

Central Region

- **Chhattisgarh, Jharkhand, Madhya Pradesh, Odisha.**

Regional Deputy Director, Wildlife Crime Control Bureau / Central Region, R.F.R.C, Mandla Road, T.F.R.I. Campus, Jabalpur- 482021.

Northern Region

- **Haryana, Himachal Pradesh, Rajasthan, Uttarakhand, Uttar Pradesh, Jammu & Kashmir, Punjab, Delhi.**

Regional Deputy Director, Wildlife Crime Control Bureau / Northern Region, Bikaner House, Shahjahan Road, New Delhi- 110 011.

In Case of Poaching of Elephants

The veterinary surgeon may be requested to find out the possible circumference of its tusk and the same may be incorporated in the Crime Scene Inspection Memo. This information is useful to correlate the tusk, if recovered subsequently, with the elephant killed.

In Cases of Killing by Poisoning

The veterinary surgeon should be requested to preserve samples of viscera for toxicological examination. The viscera samples should be sent to the concerned State Forensic Laboratory for examination by the Toxicology expert to ascertain the presence and the kind of poison used. Toxicology report should be sent to the court along with the Complaint.

- The scene of crime should be divided into smaller sectors and search for clues/evidence in those areas should be conducted sector wise.

- The evidence collected from each sector should be listed while preparing the Crime Scene Inspection Memo. The crime scene should be thoroughly searched for evidence like empty cartridges, empty bottles of poison, broken pieces of traps and tools used in the poaching and if found, the same should be collected, packed and sealed for further investigation. An area of about 500 meters encircling the scene of crime may be searched thoroughly for evidence. Generally the animal moves some distance after it is hit by the bullet or it consumed poison. It is also common that the poachers remove the carcass from scene of actual crime to a convenient location for de-skinning. Rivers, lakes or other water bodies nearby the scene of crime should also be inspected for collection of evidence as the poachers wash their body or tools used in de-skinning the animal in the nearby water bodies or rivers.

Forensic Necropsy

- Special care should be taken to preserve samples for DNA analysis. Such samples should be preserved in saturated salt solution or in silica gel (Wildlife Institute of India - WII).

- Prepare a Crime Scene Inspection Memo on the spot narrating all the events and listing all the evidence collected from the site.

- Prepare A rough sketch of the crime scene and enclose along with the Crime Scene Inspection Memo. Photograph the crime scene . All pages of Crime Scene Inspection Memo and enclosures should be signed by the Investigating Officer, and the independent witness.

- Pack, seal and label all physical objects/ evidence collected from the crime scene with the signature of the Investigating Officer and independent witnesses. If independent witness could not be associated despite reasonable efforts, it should be mentioned in the Crime Scene Inspection Memo.

- Original Crime Scene Inspection Memo along with the materials seized from the crime scene should be forwarded to the concerned Judicial Magistrate without undue delay.

- After the crime scene investigation is completed, the carcass should invariably be burnt except in case of large animals like Elephant or Rhinos where carcass may be buried. Prescribed guidelines for such disposal should be scrupulously followed.

Report Format

Name of the agency
Officer/investigator
Veterinarian's Name
Address
Contact information
Date of Examination
Subject of Examination

Accurately describe the animal – color, sex, intact or not, estimated age

Reason for Examination

Why the animal was brought
Crime Scene/Forensic Information
What the investigator told or information from investigator's report.

Medical History

Any pertinent medical history

Examination Findings

Details of your findings using medical terminology; separate subheadings are:
External Examination (weight, coat condition, body condition score, decomposition, ectoparasites, head, chest, abdomen, legs, feet).
Evidence of Medical/Surgical Intervention

Radiograph Interpretation

Internal Examination (necropsy- head, thoracic cavity, abdominal cavity, neck, respiratory tract, cardiovascular system, gastrointestinal tract, biliary tract, pancreas, spleen, adrenals, urinary tract, reproductive tract, musculoskeletal system)

Evidence of Injury (list all pertinent evidence of injury)

Procedures and Result

List all procedures, treatments, samples taken, test results or if they are pending. Additional Diagnostics: Beyond basic necropsy, additional diagnostics may be required for the investigation. Any advanced tests or analysis should be selected with great care to ensure that the maximum relevant information is obtained to draw valid conclusions, such as to:

Refute or support witness, suspect, or police reports
Establish the relationship of the animal victim to the crime scene or suspect
Establish the cause, mechanism, and manner of death and rule out other causes.
Additional diagnostic tests and analysis include histopathology, microbiology,

Forensic Necropsy

toxicology, dental analysis, entomology, DNA typing and analysis and genetic testing. Other significant testing may include the evaluation of plant and insect material found on the remains for toxicology evaluation or to determine time of death.

Forensic examination of mummified carcass may be requested, which are likely to have died under suspicious circumstances.

Mummification of carcasses can be either natural or human induced. Natural mummification occurs when a carcass is exposed to environmental factors that permit breakdown of tissues without associated microbial multiplication. Ex. In a cool dry place (in a cupboard in the house or kennel) carcasses are dry and consists of bone and desiccated tissues. Following death, bone and soft tissues may remain relatively intact in situ or they may move as a result of disturbances by humans, other animals or natural factors like rain, snow, wind earthquakes, hurricanes, waves and tides, vegetable growths, falling trees and herbal gravitational effects. The gravitational effects occur when a carcass, animal or human is on sloping ground and the remains become mingled. This may cause a disturbance in the relationship of the organs and tissues to one another.

Investigation of mummified carcasses

- Examine the environment where the animal was found including the measurement of ambient temperature, relative humidity and air flow.

- Carefully examine the exterior of the carcass to include any newspaper or substrate that is attached to it.

- Collect skin / hair samples for toxicological and microbiological/parasitological investigations.

- Rehydration in saline for 72 to 96 hrs of selected portion of skin and internal soft tissues prior to their being fixed, embedded and sectioned for histological studies

- Microbiological culture of selected tissues, either rehydrated or ground with a pestle and mortar

- Meticulous examination of bones.

Mistakes and Omissions in Forensic Necropsy

Mistakes made in forensic necropsy can have a very significant repercussions.

1. Not being aware of the objective of the forensic autopsy.

2. Performing an incomplete autopsy.

3. Permitting the body to be embalmed before performing a necropsy.

4. Regarding a mutilated or decomposed body as unsuitable for autopsy.

5. Non-recognition of changes - Bloating and discolouration, vesication, purging, nonuniform decomposition, rupture of oesophagus or stomach, autolysis of the pancreas, abnormal distension of the rectum, vulva and vagina, heat fractures, and thermal fat embolism

6. Failing to make an adequate examination and description of external abnormalities

7. Confusing the objective with the subjective sections of the protocol

8. Not examining the body at the scene of the crime

9. Substituting intuition for scientifically defensible evidence

10. Not taking an adequate photographs of the evidence

11. Not exercising good judgment in the taking or handling of specimens for toxicological examination - Unclean containers, Contamination of specimens, Permitting blood or tissue to putrefy, Inadequate samples, Poorly selected samples, Unlabelled specimens, Continuity of responsibility for protection of evidence and Facts bearing on the identity of the poison.

12. Permitting the value of the protocol to be jeopardized by minor errors.

13. Miscellaneous mistakes.

14. Mistake of talking too soon, too much or to the wrong people.

Common Errors in Necropsy

1. Performing an incomplete examination for example, failure to examine the brain

2. Inadequate documentation: for example, failure to record weights or measurements, or to take appropriate photographs

3. Recording findings too long after the necropsy

4. Failure to collect samples for ancillary analyses, or collecting samples improperly; these include inappropriate samples, unsuitable containers or preservation and inadequate labeling.

5. Accidents during the necropsy, such as contaminating tissues with ingesta

6. Mistaking changes caused by autolysis or other artifacts for significant lesions.

7. Failure to enlist the assistance of experts in other disciplines

8. Relying too much on the history and information received at the time of submission concerning the cause of death.

The Negative Necropsy

The negative necropsy is one where, after both gross examination and histopathologic examination, no conclusive cause of death is found. Reasons for a negative necropsy include:

- Failure to obtain a thorough and complete medical history and/or investigation findings

- Failure to complete a detailed external or internal examination

- Failure to complete appropriate histopathologic tissue collection

- Improper storage of remains.

However, sometimes significant external or internal signs or lesions may be absent, and the necropsy may be negative despite best efforts for correct processes. In these cases, additional toxicologic or specialized histochemical analyses may be needed to confirm a cause of death. For this reason, it is prudent to maintain duplicate tissue samples, appropriately stored for follow-up testing.

The negative necropsy doesn't indicate 'failure' since a thorough necropsy, with appropriate testing, may corroborate or contradict evidence contained in police, witness, or crime scene reports. The negative necropsy may present a significant challenge to the veterinarian's credibility as an expert witness, and therefore, must be directly addressed in any forensic report. There are now trends in the field of human necropsy for negative findings to be followed by genetic testing to identify precursor risk factors that may explain the death.

The Forensic Report

A pathology report, is not the same as the medico -legal document known as the forensic report. The goal of the forensic report is to determine the cause, manner, and mechanism of death within a reasonable degree of scientific certainty.

1. Any circumstances surrounding the death of the animal, including medical, pathologic, or toxicologic issues, should be addressed in the report.

2. The report should include pictures, supporting documents, and necropsy findings from which facts, inferences, and conclusions can be drawn.

3. All conclusions in the forensic report are formed with consideration of associated police, witness, and crime scene/investigation reports. To draw valid conclusions, alternative scenarios should be considered.

4. Finally, one should be aware that the forensic report and all evidence is subject to "discovery," meaning all legal parties will have access to the report, evidence, photographs, medical histories, and investigator notes.

5. In criminal investigation, the cause of death is sometimes referred to as the *proximate cause of death* that, *but for this event*, death would not have occurred.

6. Mechanism of death refers to the physiologic action that resulted in death (e.g. asphyxiation, hemorrhage, cardiac thrombus, renal failure).

7. Manner of death is the final determination made by the forensic investigator The legal terms used by physicians and medical examiners for manners of death, such as suicide or homicide, are not applicable in veterinary forensics In veterinary medicine, the medico-legal terms for manner of death include Natural, Accidental, Non-accidental/malicious and Undetermined.

If an animal is euthanized, the cause of death is listed as euthanasia, along with the reason for euthanasia. The manner of death is determined based on the reason for euthanasia.

5 | Examination of the Wounds

Trauma lesions are generally classified as sharp force wounds (hook or knife wound) and blunt force wounds (tower or vehicle collision). Trauma lesions may be patterned or singular. Predation and/or scavenger damage to a carcass must also be identified and differentiated from other trauma. Sharp force trauma, such as a knife wound, may be differentiated from hook trauma in illegally taken fish.

Wounds may be caused by sharp or blunt weapons, fire-arms (bullets) or by accidents by light or heavy vehicles leading to death. The death is due to haemorrhagic shock. Haemorrhages may occur in the cerebrum or body cavities (consequent to damage to the visceral organs). The presence of blood clots (big or small) in any part of the body is an indication that the haemorrhage has occurred when the animal was alive. When the injury occurs in a living animal the haemorrhage or red patch or lines of congestion are found on the skin or in the body cavities. The colour of blood clots indicates the time of injury. In a vetero-legal postmortem, it is necessary to find out the age, size, location and type of wounds including the length, breadth and depth (measure with the tape), which are mentioned, in the postmortem report. In case of firearm wounds, the diameter of the hole / wound is mentioned. Efforts should be made to search for the presence of pellets (if found, the number and diameter are mentioned in the report since these are treated as exhibits and are to be sent to the police in a sealed cover along with the postmortem report, to be submitted to

Examination of the Wounds 201

the court). It should also be ascertained whether the wounds were caused before or after the death of the animal. Mention should also be made about the probable weapon used. The injuries may be simple and grievous.

Classifications of injuries

Bruises or contusions   Abrasions   Wounds

Incised wound	Punctured wound	Lacerated wound	Firearm wound
(A cut)	(Stab wound)	(Split / tear)	

A. Bruises

Bruises are caused by a blow from a blunt weapon such as club (lathi), whip, iron bar, stone, ball, fist, boots etc or by a fall or by crushing or compression. There is no break in the continuity of the skin but there will be painful swelling and crushing or tearing of subcutaneous tissues (the integument is not broken but the underlying tissues are injured). The swelling is due to rupture of small blood vessels leading to extravasation of blood in tissues, known as Ecchymosis or effusion. When large blood vessels are ruptured, there is localized collection of blood, called haematoma. Ecchymosis appears in 1-3 hours after injury but it takes less time if the skin is thin as in eyelids and scrotum. If the ecchymosis has taken place in deeper tissues or under the fascia which is tense, it appears on the surface after an interval of 1 or 2 days or even more, at some distance from the seat of injury, following the line of least resistance and in obedience to the law of gravity. If the bruise is caused within a few hours or a day or 2 prior to the death, there may not be any ecchymosis consequent to the rapid haemolysis of stagnant blood as a part of postmortem changes. The various factors that determine the extent of ecchymosis are (1) the nature and severity of the force used, (2) vascularity of the part- in less vascular tissue the ecchymosis is very little and in more vascular tissue (eyelids, vulva, scrotum) it is extensive (3) looseness of the underlying cellular tissue (eyelids, vulva, scrotum- loose tissue- extensive ecchymosis, tough- tendons, facia- very little) and (4) the condition of assaulted victim. It may not appear in the abdomen even if a cart wheel were to pass over the body and cause death from rupture of internal organs. In case of fatal internal injuries there may not be any sign of ecchymosis on the body even when the animal is beaten with an iron tipped lathi (blunt weapon) if it is covered with a thick rug, blanket or quilt. No evidence of ecchymosis is found if the weapon used is an yielding one such as sand bag. It is a simple injury. It is seldom fatal unless accompanied by the rupture of an internal organ or by an extensive crushing of the tissues and large extravasation of blood, producing sloughing and gangrene of the parts. Several bruises may cause death from shock.

The age of the bruise can be known from the colour changes that take place due to disintegration of red blood cells and staining of haemoglobin. The changes start from 18-24 hours after infliction. The changes commence at the periphery and extend inwards to the centre. It is red at first but during the next 3 days appears blue, bluish black, brown or livid red and greenish on the 5th day to the 6th and yellow from 7th to 12 th day (as the blood disintegrates and phagocytosed). This yellow colour slowly fades till or the 14th or 15 day when the skin regains its normal appearance. Its disappearance is more rapid in healthy animals than in sickly and aged animals whose circulation is feeble. It depends on the nature of violence used. These changes are well marked and better observed in skinned animals.

Difference between ante mortem and postmortem bruises

Ante-mortem bruise	Post-mortem bruise
Certain amount of swelling and colour changes are found	No such signs
Coagulation of the effused blood in the subcutaneous tissues and infiltration of blood into the muscle fibres	No such signs

A bruise is likely to be disfigured by putrefaction and is difficult to differentiate between bruise caused ante-mortem and that caused immediately after death.

B. Abrasions

Is an injury similar to contusion but the integument is broken (the superficial layer is removed). It is produced by scraping or rubbing (slightly more severe physical force than that of contusion).

Abrasions are superficial injuries involving the loss of superficial epithelial layers of skin and produced by (1) scratch (by finger nail, pin or thorn, the objects

Examination of the Wounds

causing the scratch carries the torn epithelium in front of it and hence the direction of injury in which the injury caused is indicated by the heaped up epithelium at the end of the injury), (2) when the skin scrapes against or slides against a rough surface (graze) i.e. fall on the road and (3) by friction or pressure of ropes or strings tied around the neck or other parts of the body. Abrasions vary in size and shape and bleed very little. Abrasions due to friction against a rough surface during a fall are found in the body parts which are in contact of the ground or floor and are usually associated with contusions or lacerated wound and sometimes with very serious injuries. Abrasions are covered with mud, straw etc and their shape and patterns may indicate the type of object used to inflict injuries.

If the abrasions occur during the life, it will be covered by reddish brown crusts due to coagulation of blood / serum. The crust will be bright reddish brown in colour during the 1st 3 days but becomes dull red later. Generally abrasions heal in 10-14 days without leaving any scar. Abrasions caused after death, is dark brown parchment like in appearance due to drying and hardening of underlying skin. There is no bleeding and injection of blood vessel. From this a rough estimate of the age can be assessed.

Ante-mortem abrasion	Post-mortem abrasion
Appear as bleeding surfaces or scratches and soon covered with reddish brown crusts or scabs owing to coagulation of blood or serum. Heal in about 10 – 14 days without leaving permanent scars. If it involves whole thickness of the skin and destroys the epithelial cells capable of forming a new skin, they take a longer time to heal and leave scars (unless grafting is done).	Due to drying and hardening of the underlying skin, these are dark brown and parchment like in appearance and there is complete absence of bleeding and injection of vessels in the underlying tissues.

C. Wounds

Breach in the continuity of the soft tissues of the body including the skin, mucous membrane and cornea.

1. Incised wound

1. Produced by sharp cutting instruments such as knife, razor, sword, chopper, axe, hatchet, scythe, kookri or any object which has a sharp cutting edge and is mostly intentionally inflicted.

2. These wounds are always broader than the edge of the weapon causing it due to retraction of the divided tissues. It is somewhat spindle shaped and gaping.

3. Gaping is greater in deep wounds when the muscle fibres have been cut transversely or obliquely. Gaping is more at the center.

4. Its edges are smooth, even, clean cut and well defined and usually everted and longer than deep.

5. If a thin layer of muscle fibre is closely united to the skin, as in the scrotum, the edges may be inverted.

6. The edges may be irregular where the skin is loose or the cutting edge of the weapon is blunt as the skin will be puckered in front of the weapon before it is divided.

7. The length of the incised wound has no relation to the length of the cutting edge of the weapon.

8. The edges of a wound made by a heavy cutting weapon, such as an axe, hatchet or shovel may not be as smooth as those of a wound caused by a light cutting weapon such as a knife, razor etc and may show signs of contusion. Such a wound is associated with extensive injuries to deep underlying structures or organs.

9. A curved weapon such as scythe or sickle first produces a stab or puncture and then an incised wound and sometimes the intervening skin may be left intact.

10. It is always necessary to note its direction when describing the incised wound.

11. At the commencement it is deep and it gradually becomes shallower and tails off towards the end. But no direction is noticeable if the weapon has not been drawn while inflicting the wound

12. Haemorrhages in incised wounds are usually much more than in other wounds, since clean cut blood vessel bleeds considerably more and bleeding may be so severe as to cause death, especially if a main artery has been cut.

Examination of the Wounds

2. Punctured wound

These are called "Stabs". A stab (punctured / penetrating) wound is produced by a sharp stabbing weapon (Bayonet, dagger, knife, spears, scissors, pin, needle, pick axe, arrow) which penetrates through the skin and deeper structures. The point of instrument may be sharp or blunt. The depth of the wound is greater than the length or breadth. A punctured wound caused by a sharp pointed and cutting instrument has clean cut edges which are almost parallel but slightly curved towards each other and have sharp angles at the two extremities. This is commonly the case if the instrument has 2 cutting edges. If the instrument has one cutting edge and another blunt edge, the wound produced will show some amount of bruising and raggedness at one end of the wound. The wound usually will be wedge shaped. A sharp pointed or cylindrical or conical weapon produces a circular wound or wound with slit like opening. A blunt (pointed) instrument inflicts a punctured wound with lacerated edges (since force is needed to puncture the skin). The opening in the skin is a little smaller in length than the breadth of the weapon (due to elasticity of the skin). The depth of the wound is much larger than its length or width, or may be equal or less than the length of the blade causing it. If the instrument used is withdrawn with lateral movement the wound will be larger. In case of punctured wound perforating a part of the body, two wounds will be noticed i.e. one at the point of entry (large wound with inverted edges) and the other at the exit (smaller and with everted edges). When a stab wound penetrates into body cavity, it is called as penetrating wound. If it has gone through a part of the body and the weapon has come out from the other side , it is called as perforating wound. In some cases the depth may even be grater than the length of the blade owing to the fact that the force of the blow may depress the tissues of the part struck, allowing the point to reach the deeper tissues such as in the abdomen. There may be very little external haemorrhage and yet profuse haemorrhage may take place internally owing to the penetration of some vital organ. In some cases 2 or more punctures may be found in the soft parts with only one external orifice. This shows that the instrument has been partially withdrawn after it pierced the tissues and thrust again in a new direction. If the animal falls on a sharp pointed piece of an earthen ware pot or broken glass, the edges of the wound are irregular and more or less bruised and fragments of such articles may be found embedded in the soft tissues.

3. Lacerated wound

A wound characterized by a tearing of tissues and produced by relatively blunt objects. Lacerations occur when the legs of animals become entangled in wire and also due to automobile accidents.

A lacerated wound is (tears or splits) produced by blows with blunt objects and missiles, by violent fall on hard / sharp projecting surfaces, by wheels of vehicle, by claws, horns or teeth of animals and by projecting nails. These wounds do not correspond to the shape and size of the weapon / object causing them.

- The edges are torn, jagged, irregular and swollen or contused.

- The tissues are torn and the skin around the seat of injury is ecchymosed and the underlying bones are likely to be fractured, while the internal organs may be injured.

- Foreign bodies such as earth, grease, machine oil, cinder, hair, fibres from clothing etc are found in the wounds which predispose the animal to infection.

- When produced by blunt weapon such as club (lathi), crowbar, stone, brick etc a lacerated wound is usually accompanied by a considerable amount of bruising of the surrounding and underlying tissues and has inverted and irregular edges.

- When a heavy weight like a wheel or heavy cart or a truck passes over any of the extremities, it tears the skin and crushes the muscles and soft parts beneath it, releasing considerable blood and fat in them. Crush syndrome or fat embolism may occasionally follow.

- The direction of shelving of the margins of a lacerated wound indicates the direction of the blow applied to cause the wound.

- Haemorrhage as a rule is not extensive since arteries are not cut evenly but torn across irregularly so as to facilitates clotting of the blood.

- In the lacerated wounds of the head the temporal arteries spurt freely and forcefully when cut cleanly. These arteries being firmly bound are unable to contract and may continue to spurt and bleed for a long time.

Examination of Wound

The following information should be given

- Nature of injury: whether cut wound, bruise, burn, fracture or dislocation
- Size of each injury. Length , breadth and depth (in inches).
- Part of the body inflicted
- Simple/grievous/dangerous
- By what weapon
- Whether the weapon was dangerous or not
- Remarks

Simple injuries: Injury which is neither extensive nor serious, which heals rapidly without leaving any permanent deformity or disfiguring.

Grievous injury: Emasculation, permanent privation of the sight of either eye, hearing of either ear and of any membrane or joint, destruction or permanent impairing of the powers of any membrane or joint, permanent disfigurement of the head or face, fracture or dislocation of a bone or tooth, any hurt which endangers life or which causes the sufferer to be, during a period of twenty days, in severe bodily pain or unable to follow ordinary pursuits.

Causes of death from wounds

Immediate or Direct causes	Remote or Indirect causes
Haemorrhage Injury to a vital organ Shock	Inflammation of internal organs-such as meningitis, encephalitis, pleurisy, pneumonia, peritonitis etc. Septic infection of a wound causing septicaemia, pyaemia or exhaustion from prolonged suppuration. Gangrene or necrosis resulting from severe crushing of parts and tearing of blood vessels or crush syndrome. Thrombosis and embolism. Infective diseases such as erysipelas and tetanus. Fat embolism, Air embolism Neglect of injured animals Results of surgical operation

Immediate or direct causes

1. Haemorrhage may be internal or external or both. External haemorrhage may produce shock (fall in Blood pressure). Sudden loss of blood is more dangerous than slow loss. Loss of one third of the blood cause death. (In adults the total quantity of blood in the body is about 5%). Internal haemorrhage occurs in penetrating and gun shot wounds. Small quantity of haemorrhage in the brain or pericardium may prove fatal, blood in the wind pipe cause asphyxia and rupture of spleen, liver, lungs and heart are fatal.

2. Severe injury to vital organs (brain, heart, and lungs) is rapidly fatal. When death occurs from a slight injury inflicted on a previously diseased organ, (such as rupture of enlarged spleen, perforation of intestinal ulcer, bursting of aneurysm etc), the assailant inflicting such an injury cannot be charged for killing but can be convicted for simple or grievous hurt, if proved that his intension was not to kill the animal.

Remote or Indirect Causes

1. Death may occur due to shock without any visible injury due to paralysis of heart by a hit in the cardiac region. Death occurs easily in severely ill and weak animals, old and young animals as well. Delayed shock occurs in animal in crush injuries. Shock appears immediately after receiving the injuries but it may supervene after sometime, if the animal is in a state of great excitement at the time of receiving injuries.

2. If dyspnoea, restlessness, pericardial pain, cerebral disturbances, coma and hyperpyrexia develop some time after injury, one can suspect fat embolism. This is seen after fracture of long bones, rarely after injury to subcutaneous fat. Liquefied fat enters the blood vessels when injury to fatty tissue occurs or oil may get into the circulation when it is injected into the uterus for the purposes of abortion or investigation.

3. In injury to jugular, cephalic and femoral veins air may be sucked in because of the negative pressure during inspiration and also when intravenous injection is done (Air embolism).

4. Due to negligence of injured animal death may occur due to complication arising from simple injury.

Examination of the Wounds

5. Death may follow a surgical operation performed for the treatment of an injury.
6. If multiple injuries are inflicted on an animals by more than one person either at the same time or at different times, it is essential to identify the injury which proved fatal. Examine the wounds individually. Note which injury caused injury to vital organs or large blood vessel or led to secondary results causing death.

Postmortem findings in Death from injury/accident.

- Evidence of trauma- injury to large blood vessels.
- Extreme pallor of all tissues.
- Watery thin blood.
- Large extravasation of blood if internal injury present.
- In chronic haemorrhage, anaemia and edema.

4. Firearm injuries

Firearms are one of the major causes of death and debilitation in wildlife. The veterinary practitioner must determine (1). whether the animal was shot, (2). whether that led directly or indirectly to the animal's death or injury. A thorough examination, good radiography and a full postmortem examination will determine whether the animal was unlawfully injured or killed. The pathological examination of gunshot wounds will be very critical in determining facts around a shooting incident. Pathological evaluation is based on the wound characteristics which will assist in the investigation or prosecution of an illegal shooting.

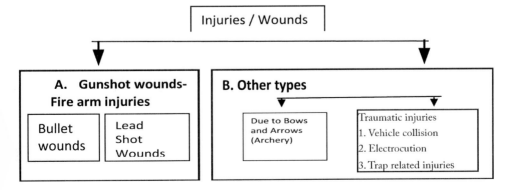

A. Gunshot wounds

Gunshot wounds are a common finding in forensic necropsy cases. In order to properly assess gunshot wounds, it is necessary to have an understanding of the weapons and ammunition that are commonly used. This information will greatly assist an investigator with analyzing the characteristics(for accurate evaluation and description of lesions) of inflicted wounds and collecting relevant information and evidence at the time of necropsy.

An enforcement officer would like to know the following:

- Type of weapon used, Number of shots and type of projectiles,

- Position of shooter relative to victim (trajectory),

- Lethality or incapacitation caused by wound and

- Duration of wound, and Recovery of projectile for evidence.

The lead "snowstorm" effect of a high velocity rifle bullet may be demonstrated on a radiograph. In conducting necropsies involving bullet wounds, it is important to determine the trajectory of the bullet into or through the body. This information may resolve the question of whether the accused hunter was properly defending him/ herself against a charging animal or illegally hunting a protected species. Basic knowledge of the types of firearms and ammunition used in hunting is essential for the evaluation of gunshot injury in wildlife. It is important to have a basic knowledge of common fire arms.

All firearms consist of metal barrel or tube of varying length which is closed at one end, called the breech end and the other open and is called the muzzle end. All projectiles are loaded at the breech end and they are forcibly driven forward along the barrel by the detonation of an explosive charge in the chamber at the closed end of the barrel. The projectile reaches its maximum velocity as it comes out at the open end of the barrel and this is called the muzzle velocity. Pulling a trigger releases a hammer or pin which by striking the percussion cap at the base of the cartridge fires the propellant charge and cause detonation. A service gun needs about 5 Lbs of pressure on the trigger, while a revolver needs only a few ounces of pressure. The breech pressure in a rifle is about 20 tons while in a pistol it is about 6 tons. There

Examination of the Wounds

are also weapons where the firing mechanism is automatic and in some even the loading of cartridges is done automatically from the containers called magazines.

Rifled fire arms

1. Those that are fired from shoulder are known as rifles. These have long barrels with bore varying from 0.22 to 0.303, a firing range up to 1000 yards (fire a single bullet and have one barrel only) and can cause serious or fatal injury and those that are fired from the hand are called pistols—light, short barreled and of different types (A revolver carries 6 cartridges in a cylindrical revolving chamber and can be fired 6 times while a self loading automatic pistol is reloaded as a result of recoil from a magazine attached to it, whose bores vary from 0.22 inches).

2. Firearms produce 2 wounds or apertures. An entrance wound and the other the exit wound. The wound entrance is usually smaller than the projectile because of the elasticity of the skin and is round when the projectile strikes the body at right angles and oval when it strikes obliquely.

3. The flame and the forceful expansion of the gases of explosion in the skin and subcutaneous tissues usually causes a large entry wound, the edges are ragged and everted.

4. Wadding or debris may be found lodged in the wound and the skin surrounding it will be scorched and tattooed with particles of unconsumed gun powder.

5. If a revolver is fired close to the skin but held at an angle, the smudging and tattooing is limited only to one side of the bullet hole. The wound of exit is often larger than the wound of entrance and its edges are irregular and everted but free from scorching and tattooing. The edges of both the wounds of entrance and exit may be evrted in fat animals due to protrusion of fat into the wounds and in decomposed bodies because of the expansible action of the gases of putrefaction. The edges of wound of exit may be very ragged and torn, if the projectile was discharged at close quarters and passed through the bone or was deformed by striking elsewhere first (wound of recoil). These characters of the wound are due to the wobble of the projectile, its deformed condition, laceration of the skin by fragments of bone expelled from the body along with the projectile or by the splintered pieces of the projectile itself.

Nature of projectile

6. Large bullets cause greater damage to the internal organs. Round bullets produce larger wounds than the conical ones. They cause extensive lacerations of the tissues and comminuted fracture of the bones if they strike the body at different angle and sometimes their course is arrested by coming in contact with chains or other hard articles.

7. The conical bullets produce much less laceration than round bullets and cause punctured wounds and rarely splitting of tissues unlike round ones..

Velocity of projectile

8. A bullet travelling at high velocity produce a clean, circular punched out aperture or slit as in a stabbing wound and usually perforates the body. It is not deflected from its path by striking a bone but may cause its comminution or splintering. A bullet of low velocity causes contusion and laceration of the margins of the wound of entrance. It is easily deflected and deformed by striking some hard object and often lodges in the body. The track made by the bullet widens as it goes deeper. This is the reverse of a punctured wound. If a bullet gazes a bone, it may produce a gutter with or without fracturing it and may or may not give the direction or deflection of the bullet.

Distance of firearm

9. If a firearm is discharged very close to the body or in actual contact with it, subcutaneous tissue over an area of 2 or 3 inches around the wound of entrance are lacerated and the surrounding skin is usually scorched and blackened by smoke and tattooed with un-burnt grains of gun powder. If the powder is smokeless, there will be no blackening of the skin but there may be a grayish or white deposit on the skin around the wound. No blackening or scorching is found if the firearm is discharged from a distance of more than 4 feet. These signs may be absent even when the weapon is pressed tightly against the body as the gases of the explosion and the flame, smoke and particles of gun powder will all follow the track of the bullet in the body.

10. The effects produced by small shot fired from a shot gun vary according to the distance of the weapon from the body and the chocking device. A charge of small

Examination of the Wounds

shot fired very close to or within a few inches of the body enters in one mass like single bullet, making a large irregular wound with scorched and contused edges and is followed by the discharge gases which greatly lacerate and rupture the deeper tissues. Particles of unburnt powder expelled from the weapon behind the missile are driven to some distance through the wound and some of them are found embedded in the wound and the surrounding skin which is also singed and blackened by the flame and smoke of combustion.

11. The exit wound of a close range shot shows greater damage to tissues than the entrance wound, the margins are everted but there is no evidence of blackening or singing. The skin surrounding the wound is blackened, scorched and tattooed with unconsumed grains of powder.

12. At a distance of 6 feet the central aperture is surrounded in an area of about 2 inches in diameter by separate openings made by a few pellets of the shot which spread out before reaching the mark. The skin surrounding the aperture is not blackened or scorched but is tattooed to some extent.

13. At a distance of 12 feet the charge of shot spreads widely and enters the body as individual pellets producing separate openings in an area of 5 to 8 inches in diameter but without causing blackening, scorching or tattooing of the surrounding skin. This scattering of shot depends upon the size of the gun, the charge of the powder and the distance of the gun from the body. As the distance increases, the damage caused by a single pellet diminishes, until at about 30 feet, it is only capable of penetrating the skin of an animal.

Time of firing

- After recent discharge—a black deposit of potassium sulphide mixed with the carbon is found in the barrel of the firearm, if black powder is used. Up to 5 to 6 hours—this deposit forms a strongly alkaline solution when dissolved in distilled water and emits an offensive odour of sulpuretted hydrogen. If the solution is filtered and the filtrate is treated with a solution of lead acetate a black precipitate of lead sulphide is formed. After exposure to air and moisture for a few days the potassium sulphide becomes converted into thiosulphate, thiocyanate and finally potassium sulphate which gives neutral solution with distilled water which gives a white precipitate with lead acetate. Later oxides of iron (iron rust) with traces of iron sulphate are formed in the barrel.

- Smokeless nitro-powders leave a dark grey deposit in the barrel of a recently discharged firearm. It does not change with lapse of time, gives a neutral solution with distilled water and contains nitrites and nitrates but not sulphides, If chromate or bichromate powder is used, the residue in the barrel is usually of a greenish tint.

- A deposit will not be found if the weapon has been thoroughly cleansed after discharge and if firearms are dirty no one can find the time of firing.

- Firearms produce two wounds, one at the point of entry and the other at the exit. If entrance wound alone is present, it indicates that the bullet is lodged in the body. In such cases the bullet has to be taken out and forwarded to the police in a sealed cover. A bullet must be carefully handled. Forceps or other metallic instruments must not be used in handling them, for, this may cause additional scratches apart from the riffling marks on the bullet.

- The wound at the entrance is smaller than the projectile and is round when it strikes the animal perpendicularly and oval if it strikes obliquely.

- The entry wound produced by a revolver fired near or in contact with the skin is stellate or cruciform in shape instead of circular. Beyond 12" distance there will not be powder mark left around the wound. When fired close to the skin but the weapon is held at an angle, the smudging and tattooing is limited to one side of bullet hole. There will not be any scorching and tattooing at the exit.

- The large (nature) bullets cause greater damage to the internal organs. The bullet traveling (velocity) at high speed produces a clean circular punched out opening or slit as a stabbing wound and a bullet traveling at a low speed causes, contusion and laceration at the entrance.

- If the firearm is fired from a close range (distance), the subcutaneous tissue over an area of 2-3" round the wound of entrance is lacerated, surrounding skin is scorched and blackened by smoke and tattooed with un-burnt grains of gun powder. If the powder is smokeless, (nitrocellulose) no blackening of skin occurs but there may be grayish or white deposit on the skin around the wound.

- If fired from a distance of 4' or more, no blackening and scorching noticed.

- The rifling gives the bullet a spin, greater power of penetration, straight trajectory and prevents it from wobbling as it travels in the air. Empty casings of revolvers have a rim at their base while those of rifles and pistols do not have a rim. The

Examination of the Wounds 215

size of the casing provides a clue regarding the caliber of the weapon used. Rifled weapons employ cartridges with smokeless powder which does not cause much blackening of the skin because it burns up entirely while the black powder of short guns causes blackening of the skin. The revolver bullets are propelled at a velocity of 500 feet per second, the pistol bullets at 1000 feet per second and the rifle bullets at 2500 feet per second. Since the damage inflicted depends on the velocity of the bullets, a rifle is more deadly than the other two weapons.

Examination of firearm injuries

Radiograph examination is invaluable in detecting injury caused by air rifle pellets, shotgun pellets, and rifle bullets. The type of projectile and the direction of travel through the body can be observed. Finding the projectile or major fragments, if present, is enhanced by the x-ray. The bullet jacket, important in the matching of the bullet back to an individual gun, may be differentiated from the bullet's lead core which has little use in comparison analysis. Bone fractures caused by the projectile can also be assessed. Not all bullets leave fragments in the wound (Ex: all copper Barnes bullets, military full metal jacket bullets).Radiographs are taken before the full postmortem examination begins so that pellets or bullet fragments can be located and recovered. Pellets may be lying superficially in the skin or feathers or may be lodged under the skin and are easily lost if care is not exercised. Properly bag the specimens in order to prevent loose shot and pellets being lost and to avoid allegations of cross-contamination when stored with other specimens which may also have been shot. The finding of shot in an animal may not necessarily relate to the cause of death or injury. Many wild birds and mammals suffer non-fatal shooting injuries and it is vital that the forensic practitioner determines whether any bullet fragments or pellets are related to the cause of death or current injury, or whether they are the result of a previous incident. For example, lead shot may be found in the body, where it is not interfering with any vital functions and so may be of no clinical significance. The presence of shot may also raise concerns about the possible use of poisons. It is important to establish exactly where the pellet is located in the body so that the presence of incidental lead shot can be eliminated. Many wildfowl may carry shot in the tissues from previous shooting incidents or pick it up whilst feeding. This lead is often ground into thin discoid objects and may be recognizable on a radiograph. Birds of prey may ingest lead when scavenging on shot animals and this can remain in the gizzard for long periods of time, sometimes leading to lead poisoning.

i. Wound features

- Gunshot wounds caused by small calibre weapons may not be evident immediately on external examination. .

- Detecting bullet wounds in non-human species presents problems due to the presence of hair and feathers.

- Some gunshot wounds bleed little and virtually are undetectable during gross examination. Hence, it is necessary to skin the entire carcass to document all of the wounds.

- When a victim is shot at close range the skin surrounding an entrance wound shows "tattoo"(staining of skin by gunpowder that is lodged in skin). In animals, most wounds are inflicted from a distance, with the exception of cases where a final kill shot is administered.

- An abrasion ring at the entrance wound with breaking of adjacent hair or feathers is seen due to injury by blunt projectiles, such as slugs or shot. This lesion may not occur with a bullet. The "typical" bullet entrance wound is characterized by a smooth hole in the epidermis / dermis.

- Radiographs can provide information on the trajectory of wound. On radiographs, a high velocity bullet may result in what is known as a "lead shower", whereby the bullet fragments as it penetrates deeper into the wound. The result of this fragmentation is a conical shaped wound track.

- Shotgun pellets, slugs, and muzzleloader balls do not typically result in a lead shower and remain intact. The outcome is a more cylindrical wound track as demonstrated by similarly sized entrance and exit wounds.

- Another use for radiographs is preliminary differentiation of shotgun pellet type.

- Lead shot is more likely to be deformed during collision than steel or alloy shot, thus radiographs can suggest the make-up of the shotgun pellet used.

- However, chemical analysis is necessary to confirm the use of lead pellets.

- Note: One should never attempt to determine calibre or ammunition size based on the radiographic appearance as this method is extremely inaccurate.

Examination of the Wounds

ii. Finding Projectiles

- In many cases it may be necessary to locate and remove shot or bullets and fragments for identification or submission for further forensic examination. Using radiographs taken in two planes can help locate pellets or bullets prior to dissection.

- Metal detector or whole body radiographs or fluoroscopy may be used in locating metal fragments. However, even with the benefit of radiographs, locating small fragments of metal in a large carcass can be an extremely difficult.

- Skinning the carcass is vital to locating projectiles. As a projectile passes through a body, its kinetic energy becomes diminished. The elastic nature of the skin often catches projectiles and leaves them lodged in subcutaneous tissues, even in relatively thin-skinned species such as avian.

iii. Collection of Projectiles

Whole or fragmented bullets may have rifling marks from the weapon they were fired and these need to be removed and handled with care. Proper collection and preservation is vital to preserving the valuable information that can be obtained from projectiles. Improper handling can deface or obscure the striations present on bullets, thus making rifling mark comparisons difficult or impossible. Projectiles should not be collected with metal instruments. The projectiles from the carcass are important additional evidence and that a "chain of custody" must be established for the new item. Careful dissection in the area of the object should allow the pathologist to feel the object. Contact between a scalpel or other metal instruments such as forceps, which may mark the object surface, should be avoided. The item can be removed with gloved fingers or rubber-tipped forceps or gauze covered forceps. Preservation of the detail on the bullets exterior surface is essential for matching. Therefore, the bullet must be handled in a manner which does not scratch or mar the surface and must be carefully cleaned, dried and stored in a manner which prevents corrosion of the metallic surface. Fluids or tissue left on metal can lead to corrosion and deface evidence. Use of porous paper envelopes instead of plastic bags or sealed bottles will reduce the possibility for oxidation and corrosion. Wrapping the bullet fragment in gauze will protect it from possible physical damage. Following air drying, these are normally wrapped in tissue (not cotton wool) and put in an appropriate rigid container to prevent damage, friction or crushing by movement during transport.

The marks analyzed by forensic firearm experts are incredibly small and easily defaced. Comparison of bullets requires the use of a special microscope that allows the side-by-side comparison of bullets and cartridge casings. Radiographs may also be of considerable value when examining shot animals later prepared as taxidermy specimens. These may show traumatic injuries associated with the remaining skeletal structure and assist in the recovery of shot. Handling of collected projectiles should be kept to a minimum. The bullets and fragment should be gently rinsed in cold water to remove blood and tissue, and dried as any dampness may cause oxidation of the surface and hinder further examination. Pathologists may dip them in 70% alcohol to remove any water residues and to disinfect them.

iv. Proving Lethality

Gunshot wounds often are not immediately fatal. An animal (or human) shot directly through the heart can survive a surprisingly long time before succumbing to cerebral anoxia. This is relevant to forensic cases when a wounded animal, such as a deer, can travel a significant distance before death. A defendant could claim that the wounded animal wandered into his or her vicinity before dying. It would be up to the forensic investigator to analyze the evidence available and render a professional opinion as to the plausibility of different scenarios. The lethality or the incapacitation ability of the wound is important for the pathologist to document based on the damage to vital organs or structures because the animal's ability to move after it has been shot may be critical in presenting the facts of the case.

v. Bullet Wound

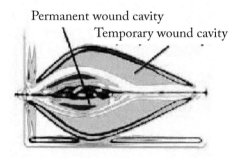

- Gunshot wounds in tissue consist of the primary projectile path (permanent cavity) surrounded by "satellite" paths from bullet and bone fragments and

Examination of the Wounds

tissue expansion and tearing trauma. High velocity projectiles significantly increase the amount of tissue damage compared to low velocity projectiles. The lead "snowstorm" effect of a high velocity rifle bullet may be demonstrated on a radiograph. The visible tissue damage to the carcass is directly correlated with the snowstorm of radio dense particles except where particles of the bullet are washed by free blood around the internal body cavities.

- Gunshot wounds generate temporary and permanent wound cavities. The size of these cavities is primarily influenced by the density and character of the tissue affected, the size and construction of the bullet and speed of impact.

- The x-axis represents depth of penetration. The wound entrance is on the left as represented by the drawing of a bullet. The outer region represents the temporary wound cavity caused by bellowing outward of soft tissue in response to impact of the bullet. This phenomena occurs at the instant of impact as a shock wave disseminates through the body. The inner region is the permanent wound cavity left by tissue destruction. The amount of damage will depend on a number of factors, including projectile mass, projectile speed and ability of the affected tissues to expand.

- Permanent wound cavities are relatively larger in tissues that are less expansile, such as bone, muscle bundles and encapsulated organs. For example, a permanent wound cavity left by a high velocity, large calibre hunting rifle will be much greater than that of a smaller 0.22 calibre rifle, which fires a smaller bullet at lower speed. Likewise, the cavity is potentially much greater when left by a bullet that fragments and greatly deforms on impact (such as hollow point) than a solid projectile that simply passes through the animal.

- The character of a tissue affects the ability of a system to tolerate the force generating the temporary cavity. For example, hollow viscous organs (bowel) are more expansile and simply may be perforated, while a kidney is encapsulated (thus preventing expansion) and would be obliterated by the same projectile. Bones may fragment, creating a shower of bone shards that function as secondary projectiles.

- Blunt projectiles, such as slugs or shot may leave an abrasion ring at the entrance wound with breaking of adjacent hair or feathers. This lesion may not occur with a bullet. The "typical" bullet entrance wound is characterized by a smooth hole in the epidermis/dermis.

- The different outcome could be due to differences in impact speed, bullet mass, or design of the bullet used.

A bullet produces tissue damage in three ways.

Laceration and crushing	Cavitation	Generation of shock waves
A bullet can shred (lacerate) or crush tissue or bone. Bullets moving at relatively low velocity do most of their damage this way. Fragmentation of bone can cause further damage, as the bone shards themselves become missiles. Low velocity bullets, as in handguns, that travel less than 1000 fps do virtually all their damage via crushing.	A "permanent" cavity is caused by the path of the bullet itself, whereas a "temporary" cavity is formed by continued forward acceleration of the medium (air or tissue) in the wake of the bullet, causing the wound cavity to be stretched outward. Cavitation is significant with projectile traveling in excess of 1000 fps. This damage is produced by the forward movement of air or tissue in the wake of the bullet. The wound that is produced by the bullet is destructively broadened by the force of the moving air or tissue. In a tissue, this produces even more structural damage.	The air at the front and sides of a very fast moving bullet can become compressed. The explosive relaxation of the compression generates a damaging shock wave that can be several hundred atmospheres in pressure. At high velocity, generated shock waves can reach up to 200 atmospheres of pressure. Fluid-filled organs such as the bladder, heart, and bowel can be burst by the pressure. Shock waves compress the medium and travel ahead of the bullet, as well as to the sides, but these waves last only a few microseconds and do not cause profound destruction at low velocity. However, bone fracture from cavitation is an extremely rare event.

Since the damage inflicted depends on the velocity of the bullet, a rifle is more deadly than the other two weapons (the pistol and the revolver).

Examination of the Wounds

vi. Patterns of Tissue Injury in gun shot wounds

Classification

One of the commonest determinations of the forensic pathologist is the range of fire. Gunshot wounds are typically classified as

- Contact wounds
- Intermediate range or close range wounds
- Distant range wounds

The wound produced by the bullet or shot as it enters the body called the wound of entrance (entrance wound) or entry and that by which it leaves the body is wound of exit.

Classification of entrance wounds

Contact wounds	Intermediate, or close-range, wounds	Distant range wounds
1. Have soot on the outside of the skin and muzzle imprint, or laceration of the skin from effects of gases. 2. Contact wounds of air guns usually lack these features.	1. May show a wide zone of powder stippling, but lack a muzzle imprint and laceration. The area of powder stippling will depend upon the distance from the muzzle	1. Lack powder stippling and usually exhibit a hole roughly the calibre of the projectile fired.

The most difficult problem is distinguishing the distance from a contact wound. The factors that can affect the amount and distribution of gunshot residue (GSR) on skin and clothing include
Firing distance,
Length and diameter of the firearm barrel,
Characteristics of the gunpowder,
Angle between the firearm barrel and target,
Characteristics of the cartridge,
The environment (moisture, wind, heat),

Type of clothing,

Intermediate targets, and

Characteristics of the target (tissue type, putrefaction, blood marks).

Examination for Gunshot residues (GSR) may aid in distinguishing entrance from exit wounds

Differences between entrance and exit wounds

Entrance wounds	Exit wounds
The entrance wounds have smaller diameter (because the velocity of the missile burrows through the skin like a drill).	Exit wounds are generally larger than entrance wounds, due to the fact that the bullet has expanded or tumbled on its axis (during the exit, the missile tears the tissues out). Most bullets are designed to hit the target without exiting, In many situations an exit wound will be present.
The edges are inverted	The edges are everted
The entrance wound will have more GSR than the exit wound. Residue is lacking in entrance wounds with air-guns. The alizarin red S stain can be utilized in microscopic tissue sections to determine the presence of barium as part of GSR	The exit wounds do not have GSR. Exit wounds either do not exhibit gunshot residues or far less residues than associated entrance wounds. In bone, typical "beveling" may be present that is oriented away from the entrance wound.
Scanning electron microscopy of entrance wounds shows gunshot residue within collagen fibrils. The entrance wound appears abraded with loss of the papillary pattern and laceration of basement membrane. (Computer assisted image analysis may aid detection of GSR).	Scanning electron microscopy of exit wounds shows irregular lacerations with protruding collagen fibers, but relatively undamaged papillae. The edges are not abraded.

Examination of the Wounds

Entrance wounds	**Exit wounds**
Entrance wounds into skull bone typically produces beveling or coning, of the bone at the surface away from the weapon on the inner table. In thin areas such as the temple, this may not be observed. Sternum, iliac crest, scapula, or rib may show similar features. (These observations may permit determination of the direction of fire. The direction of fire of a graze gunshot wound of the skin surface can be determined by careful examination of the so-called skin tags located along the lateral margins of the graze wound trough, by use of a dissecting microscope or hand lens. Characteristically, the side of the tag demonstrating a laceration is the side of the projection toward the weapon.)	Fragmentation of the bullet may produce secondary missiles, one or more of which may have exit wounds. The bullet path may be altered by striking bone or other firm tissues, such that the bullet track may not be linear, and exit wounds may not appear directly opposite entrance wounds. (It is important to remember that the orientation of the bullet track may be positional. The victim may have been shot while standing or sitting, but when the body is typically examined at autopsy, it is lying down, so that soft tissues may shift position. This must be remembered when rendering opinions as to the angle, or direction, of fire).
"Shoring" of entrance wounds can occur when firm material is pressed against the skin, such as when a victim is shot through a wooden, glass or metal door while pressing against it to prevent entry of an assailant. Such wounds have a greater wound diameter and demonstrate greater marginal abrasion than control wounds produced by the same weapons. The features were directly proportional to the Kinetic energy (KE)of the projectile and the rigidity of the shoring material. Stellate radiating lacerations of some shored wounds could lead to misinterpretation of distant range of fire as a contact wound.	If the exit wound is "shored" or abutted by a firm support such as clothing, furniture, or building materials, then the exit wound may take on appearances of an entrance wound, such as a circular defect with an abraded margin. This can occur with contact, close range, or distant shots. Shored exit wounds have a round or ovoid defect and all have some degree of abrasion. The degree of shoring abrasion increased directly with the KE of the projectile and the rigidity of the shoring material.

Entrance wounds	Exit wounds
7. Tangential entrance wounds into bone may produce "keyhole" defects with entrance and exit side-by-side, so that the arrangement of beveling can be used to determine the direction of fire.	7. A keyhole lesion, typically identified with entrance wounds, has been described with an exit wound

Entrance wounds

- Use of silencers (or "muzzle brakes" to deflect gas and recoil) may produce atypical entrance wounds. A silencer is a device, often homemade, fitted over the muzzle that attempts to reduce noise by baffling the rapid escape of gases. Their possession is illegal.

- Entrance wounds produced when silencers are present lead to muzzle imprints that are erythematous rather than abraded and disproportionately large for the size of the wound. Entrance wounds may appear atypical at close range.

- Firearm missile emboli ("wandering bullets") are rare, but may occur in victims that survive for some time and may require surgical intervention.

- Entrance wounds associated with black powder handguns are associated with extensive sooting, a long range of travel of the sooting into the wound and skin burns. Large pocket-like underminings may be seen even in deeper tissue layers with contact range wounds.

- The skin defect at the entrance site occurs from multiple mechanisms. Most of the defect results from skin fragmentation with fragments carried into the bullet track. The negative pressure of temporary cavitation pulls skin particles into the wound. There is also back spatter of skin particles away from the direction of bullet travel upon entrance.

- Infection may result from gunshot wounds. Bullets are not sterile objects, either before or after firing. Bacteria are ubiquitous on skin surfaces and clothing. The bullet carries bacteria into the wound track. Skin particles serve as a transport vehicle for the bacteria.

- The general principle that the exit wound is larger than the entrance wound may not apply to smaller ammunition that passes through the body without

Examination of the Wounds

significant deformation such as some 0.22 caliber ammunition or shot pellets and slugs. As high velocity rifle ammunition fragments on impact, it may leave a much larger exit wound with extensive damage and exteriorization of tissue. Also, projectiles that fragment within a body may exit in multiple locations. Another technique that can provide information on the trajectory of wound is radiography. On radiographs, a high velocity bullet may result in what is known as a "lead shower", whereby the bullet fragments as it penetrates deeper into the wound. The result of this fragmentation is a conical shaped wound track. Shotgun pellets, slugs, and muzzleloader balls do not typically result in a lead shower and remain intact. The outcome is a more cylindrical wound track as demonstrated by similarly sized entrance and exit wounds

vii. Sequence of fire

- In some situations, pathologic findings may help to establish in what sequence the bullets were fired that caused the injuries. For example, multiple gunshot wounds to the head may produce fracture lines, and a subsequent fracture line will not cross a pre-existing fracture line

- Subjective reasoning would suggest that the first shot may be horizontal (victim upright) but subsequent shots would be oriented down or to the back of the victim as he fell or fled. Without witnesses and scene investigation, such opinions would be conjectural.

- The management of gunshot wounds may require accounting for all bullets and bullet fragments to determine the need for surgery. A simple rule of accounting for bullets is as follows: the number of entrance wounds must equal the number of exit wounds plus bullets retained. An unequal number may result from bullet fragmentation or from embolization, migration, or ricochet to unsuspected tissue sites. Radiographic imaging may be needed to account for retained bullets and fragments.

viii. Issues in Wildlife Gunshot Cases

Lead Shot Pellets

1. Identifying lead shot in a carcass is simple by chemical methods.

2. The story becomes confused by the fact that game species have international flight paths and that non-lethal, minimally debilitating wounds can occur during migration.

3. Questions that require answers include

   Is the defendant responsible for the lethal wounding?

   Is the illegal type of ammunition responsible for this wound?

   Could the lead shot be from an incident prior to the defendant's inflicted wound?

4. Answering these questions would require

   Assessing the lethality of the wound,

   The age of different wounds

   The type of shot involved with each.

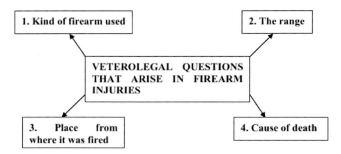

1. Kind of firearm used can be deduced from

An examination of the bullet i.e. its size, the number, size and the direction of rifling marks on it and the kind of metal of which it is made—all these will give clues regarding the weapon used. Ex. A 0.303 bullet indicates that it has been fired from a weapon of a caliber of 0.303 inches. The bullets can be recovered from the scene of offence or from the dead animal during necropsy. The bullet must be carefully handled.

Examination of the Wounds

The suspected weapon, the cartridge and the bullet must be examined by a ballistic or fire-arm expert.

Observations	Probable Firearm used
Recovery of lead shots, glass beads or stones.	Shot guns
Wad at the scene of offence or in the dead body. The diameter of the wade may give a clue regarding the caliber of the weapon.	
Presence of wad	Shot gun
No trace of wad	Rifled weapon
Empty casings.	
The casing is the outer shell of the cartridge which encloses within itself the powder, shots or bullet and wad. When these have been duly discharged after detonation, the remaining shell is called the empty casing. The size of the casing provides a clue regarding the caliber of the weapon used.	
Rim at base of the empty casings	Revolver
No rim	Rifle and pistol
Stains on the body of the animal	
Not much of blackening of the skin (cartridges with smokeless powder burns up entirely	Rifle
Blackening of the skin (due to black powder)	Shot guns
5. The appearance of the wounds caused by the firearm also helps in determining the weapon used.	

2. Shooting distance

In most cases involving animals the weapon is discharged at a mid to distant range and there is no evidence of tissue damage related to heat, expanding gases and propellant residues. When shotgun injuries are found the pathologist is asked to estimate the discharge distance. Shot of the same size can be fired from shotguns of varying calibre which can produce very different shot patterns from the same distance. Knowledge of the weapon used may assist with any assessment. In birds and small animals, it is virtually impossible to give an accurate estimate unless the gun was discharged less than a few metres from the victim. At greater distances, the mass of shotgun pellets diverges to a degree that causes many of the pellets to miss the victim. Consequently, the full size of the pattern of shot cannot be determined and the discharge distance remains a matter of conjecture.

The range or the distance from which the weapon was fired is deduced from the

- Size and nature of the wound,
- Blackening,
- Burning and
- Tattooing.

The barrel is generally lubricated between use. The missile as it is propelled through the barrel carries this lubricant (oil) on it which may be deposited on the skin around the central hole. This is called the grease collar or the dirt collar. When a weapon is discharged the elements that come out of the muzzle are the flame, missile and smoke (un-burnt powder). Of these the missile travels longest and the flame the shortest distance from the muzzles. If this is remembered, it is easy to visualize the nature of the injuries that may be caused by the discharge of fire-arms from various distances.

Types of gunshot injury

Three types of firearms may be recognized through the use of radiography, examination of the wound and the type of shot. These are shotguns, air rifles, and rifles (rim fire and centre fire). The first two are the most frequently encountered.

Shot gun injuries

Shotgun pellet wounds are often multiple. The number and concentration of pellets is dependant on the size of the pellets and the distance from the shooter. Shotgun wounds containing very high concentrations of pellets indicate very short range gunshot and should be examined for the presence of plastic or fiber wad material which are not radio dense and do not show up on x-rays. These wads are important evidence items as they indicate the gauge of shotgun used. Shotguns fire a number of round pellets at speeds of approximately 500m/sec. The penetration and the damage done to the tissue is dependent upon the number, size and speed of the pellets as they

Examination of the Wounds 229

hit the animal. As the pellets are of a relatively low velocity and mass, there is a good chance that they will be retained in the body. Shotguns are relatively short range, and so in most cases, but not all, multiple pellets are seen on the radiograph. However there is a significant variation in pellet spread which is dependent on distance, the choke used in the barrel(s) and the differing bore of shotguns as well as the type of shot (which varies in size and the number in a particular cartridge). Consequently, the number of pellets seen in the radiograph cannot be used to accurately define the distance to target, or the bore of shotgun used. Lead is the most common shot pellet component, however pellets may also be made of copper- coated lead, steel, bismuth, and even tungsten. Pellets that deform, or produce fragments are typically made of lead or copper coated lead, though bismuth and tungsten will also deform in this manner. In certain locations and for certain species the use of lead shot may be prohibited. As steel pellets remain intact and round, even when hitting bone, a radiograph may give an indication whether lead shot has been used.

Appearances found	Probable range of fire
A large hole (due to the explosive force of the gases), edges of the hole are seared by the flame and the skin surrounding it is blackened by smoke (This blackening is also called smudging).	Weapon is fired with the muzzle held in contact with the skin.
Central, fairly large and irregular hole (due to the shots entering the animal in one mass). The skin around this hole is seared by flame and smudged by smoke. Around this is a zone of tattooing or pepper pattern produced by the grains of gun powder being driven into the skin. This tattooing is seen as a small discrete black specks.	Weapon is discharged from a little distance from the animal
No burns and smudging and tattooing over a wider area.	Weapon is discharged from a distance greater than that to which the flame travels.

Appearances found	Probable range of fire
Only small discrete holes caused by the individual shots	Weapon is discharged from a greater distance than that to which the flame travels
Small discrete holes (caused by individual shots) over a wider area.	Weapon is discharged from a greater distance than the above

Rifled weapon injuries

The velocity of the projectile (bullet) is extremely important in the wound characteristics. The energy available for tissue disruption and penetration increases exponentially with velocity. The mass and the construction of the projectile may also influence the wound characteristics but to a lesser extent than the velocity.

Low velocity gunshot wounds

These are caused by bullets such as from pellet guns, e.g. air rifles and 0.22 rim fire. The most common finding is an air rifle pellet. These may pass through the body but often have insufficient energy to escape. The radiographic characteristics often show what type of bullet is lodged in the carcass, e.g. air rifle pellets, or occasionally very small bullet fragments, usually associated with a damaged bone. The wound path is characterized by a long single relatively narrow channel, with few or no metallic fragments associated. The tissues may be crushed or torn, but rupturing of organs is generally not observed.

High velocity gunshot wounds

High velocity bullets tear and shred tissue in a much larger radius around the path of the projectile. Most hunting is done with either hollow nosed or soft nosed bullets which are designed to deform in a predictable manner. This type of expanding ammunition increases the rate at which the energy transfer takes place and will create larger and more predictable permanent wound cavities than solid ammunition. The damage done to the tissue will vary depending on the type of firearm, the ammunition and the shooting distance. The entrance wound is normally small, with large, often multiple, exit wounds, depending on whether the bullet has mushroomed or broken up inside the animal. High velocity bullets will often penetrate through the animal and then be spent in the environment. However, occasionally spent pieces of the

Examination of the Wounds

bullet will be found opposite the entrance wound in the subcutaneous tissue, this is often comprised of pieces of the outer jacket of the projectile, and can give rise to confusion with shotgun fragments on radiographs.

Appearances found	Probable range of fire
Stellate or star shaped, imprint of the muzzle on the skin (imprint impression), with burning and some blackening, but no tattooing.	A contact injury
Round hole with abraded edges (called abrasion collar), some contusion with the abrasion collar (called the contusion collar) Grease collar in the inner zone, with some blackening and tattooing	From a distance of about a metre
A punched out hole with no blackening, tattooing or singing	From a distance of a metre or more

1. The range is established accurately by firing test shots at different ranges by using the same weapon and similar cartridges. This is done by the ballistic expert.

2. The place from where it was fired is deduced from the following data

i. The direction of the injury - If the line joining the wounds of exit and entry or the wound of entry and the site where the projectile is lodged is projected forward, it generally indicates the direction of fire unless, the projectile has been deflected in its course within the body as may happen when it strikes bone.

If the bullet enters through the skin perpendicularly the abrasion collar is uniform with around the central hole, while if it enters at an angle, the abrasion collar is wider on the side from which it entered. This also indicates the direction of fire.

ii. The range is determined from the appearance of the wound. - Since the direction and distance are now known, the place from where the weapon was discharged can be determined by projecting the direction of fire in a straight line.

iii. The cause of death from fire arm injuries may be from

- Injury to a vital organ
- Shock and haemorrhage or \septic complications.
- Toxic shot (lead shot) must be differentiated from non toxic shot in waterfowl legal cases. It is essential to document with pathological observation and shot recovery and analysis if toxic shot caused death or incapacitation of the waterfowl. Projectiles recovered from wounds are valuable forensic evidence. Bullets may be matched to an individual gun by a firearms examiner.

B. Other types of injuries

1. Archery

Modern bows are capable of firing arrows that can travel completely through the body of adult cervids. Arrows used are usually tipped with broad-heads, which are steel points hedged with cutting blades. The wounds inflicted by arrows have a few characteristics that distinguish them from those of firearms. The wounds are made by incision, thus the skin edges may be re-apposed for almost complete closure of the wound in a fresh hide. In contrast, firearm wounds usually obliterate the epidermis/dermis upon impact and may leave an abrasion ring on the deep surface of the hide. The wound tract caused by an arrow is relatively narrow and linear.

It is necessary to know these characteristics when unscrupulous hunters kill an animal with a firearm and then punch an arrow through the wound to mimic an archery kill. Field tests are available for testing tissue for lead and copper residues, as would be left in a gunshot wound. Unfortunately, these tests are not very sensitive and additional analyses must be performed in the laboratory.

In addition to the above distinguishing features, radiographs can be used to look for metal fragments. Evidence of metal fragments or a "lead shower" is strongly indicative of a gunshot wound as broad-heads do not fragment. Differentiation of arrow wounds from gunshot wounds into which an arrow has been placed is a common law enforcement request. Pathological examination, radiographic documentation and finally lead analysis may be required to prove that a bullet has made the original wound. In a whole carcass examination, it is also important to document that the wound is incapacitating and/or lethal.

Examination of the Wounds

Bullet wounds	**Wounds caused by Arrows**
Mechanism of wounding	
Most bullets(from pistols and .22 rim fire rifles) travel at velocities 750 fps to 3500 fps (foot per second)	Arrows travel at velocities around 350 fps
Crush or tear tissues along the wound tract. As bullet enters, kinetic energy is lost into the tissues. The higher the velocity and or mass of the projectile, the greater the energy and the more potential for radiating damage. Fragmentation of the bullet associated with mushrooming and penetration and fragmentation of the bone mass add to the severity of the wound by providing additional projectiles which tear the tissues along the wound tract. The tissues which have been torn or crushed by this process get filled with blood from the disrupted small blood vessels in the area. Highly elastic tissue like lung may absorb the kinetic energy and the crushing effects of a bullet in a manner very different from less elastic tissue like bone, muscle or liver which may fracture or tear giving a more pulverized appearance.	Arrows are designed to cut vital organs. They will not cut or tear tissues along the wound tract. Arrow wounds are fundamentally different and should be kept in mind while examining the carcasses which are claimed to be archery killed animals. The arrows broad head cuts its way through the tissues. Mass of the arrow is relatively large but the velocity is slow compared to the bullet. Arrows do not mushroom or fragment on impact so satellite projectiles are not produced. Bone may be fractured but transfer of velocity from the arrow to bone fragments does not occur as in bullet. The arrow cuts its way through tissues leaving very limited tissue disruption other than cutting of tissue along the actual wound path. Haemorrhage from incised large vessel or organs (such as heart) leads to death from a well placed arrow in a clean kill. The tissue surrounding the arrow wound track may be relatively blood-less if the blood pressure is quickly decreased from arrow penetration of the heart or aorta which is the main target of a good archer. Arrow may fracture ribs and shoulder blades or even larger leg bones. The arrow bone break is relatively clear and the wound channel does not contain small fragments of "granular" bone which are characteristic of bullet bone fracture.

Bullet wounds	**Wounds caused by Arrows**
A high velocity bullet passing through the skin may cause significant tearing of subcutaneous tissue with extensive blood infiltration. As bullet penetrates the skin it starts to mushroom or destabilize and tumble through the underlying tissue layers. In a bullet wound on the skin, bullet wipe (sooting) from the bullet will be present in the edges of the wound.	Arrows passing through the skin and into the underlying tissues will leave a stab wound with comparably little subcutaneous haemorrhage. Arrows do not have the capability to shatter the spine of a deer or elk like a bullet. Arrows may penetrate the spinal canal and severe spinal cord but capability to fracture the base of the vertebra is improbable.

Examination of evidence: Field investigations may encounter evidence in a variety of forms ranging from intact carcasses to pieces of skin or tissues. Examination of a field dressed carcasses is easier to evaluate accurately rather a capped out or skinned big game animal and a freezer full meat.

1) Intact field dressed carcass: It is very easy to miss a small caliber bullet wound inside the neck and head in a deer carcass but may be the cause of death or debilitation in an illegal kill. Skinning or X ray may be necessary to determine if additional wounds are present. Wounds in the chest, brisket area or abdomen might be further examined if they are claimed as the fatal arrow wounds. A good hunter will take a good clear lateral chest shot. Shots to neck, brisket or abdomen are unlikely in a clear kill. Check the trajectory of wound path. Examine the entrance and exit wounds to determine if a clear path through vital organs is possible and that both the wounds are of incised type. Arrows are more likely to be shoved into bullet exit wounds since these are more obvious than the entrance wounds in a heavily haired animal. Wounds through rib cage should be examined from inside the chest cavity to find out whether the wound through the chest have an incised edge with a pattern of a broad head or is there an evidence of a more extensive damage to the tissues surrounding the wound path. If there is an extensive subcutaneous haemorrhage around the entrance wound one must suspect. Carcasses of large game animals with wounds which fracture vertebra that are claimed as archery kills are also suspect.

2) Skin, parts or gut piles: Documentation of the cause of death due to legal archery kill may be difficult to know when the carcass is disassembled. The problem arises in ruling out gunshot wounds in missing parts or altered parts. However the available

Examination of the Wounds

wound tissue may be collected and examined. The cape or skin is a good place to start in a trophy animal as this is often available. Examine the characteristics of the hole in the skin. The hole should conform to the expected characteristics of cutting edges of a 3 or 4 bladed arrow head. The arrow head entering the skin at right angle may give a pattern different from a straight perpendicular entrance. Examine whether the hole represents an entrance or exit wound. The entrance wound of an arrow should have at-least some hairs that overlay the wound cut off evenly along the same place as the skin incision. A postmortem arrow wound made by some one pushing an arrow into a carcass or into preexisting bullet wound may not cut the overlying hair in an even manner because the hairs are pushed aside opposed to having been cut cleanly by the fast moving arrow blade. An arrow exit wound with the cutting edges of the arrow head emerging from the flesh side of the skin is not likely to cut hairs over lying the wound. Haemorrhages along the edges and blood caked on the hair around the wound is a good indication that the wound was a pre -mortem wound. Wound in the skin that are suspect bullet wound should be collected , protected from contamination and further examined in a forensic laboratory. In an illegal kill it is likely that the blood shot wound tissue has been trimmed from the carcass and may or may not be available for analysis. The tissues around a suspect exit wounds are more productive of metallic particles from a bullet than skin tissues around an entrance wound. X rays help in demonstrating the lead or cupper particles in the wound tissues rather than the chemical tests with sodium rhodizonate (gives 50% false negative results). It is advisable to collect particles identified by X rays from the skin and have them chemically analyzed. This may be difficult. Alternatively the wound tissue may be analyzed for excess lead residues (for this collect tissues far away from the suspected wound area from the same animal to be used as negative control). Hunting arrow points do not leave lead or copper residues in the tissues. Similarly the bullets made of solid copper or full metal jacket military type bullets also leave no lead residues in the wound tract. Histopathological evaluation of the tissues of the wound may be required to actually prove that the wound is fresh.

Many causes of traumatic injury (such as vehicle collisions, electrocution, sharp instruments, bludgeoning, and various trapping techniques). are encountered in wildlife forensics.

i. Vehicle Collisions

- Wildlife frequently fall victim to vehicle collisions. Vehicular or radio tower collisions are frequent occurrences in wildlife. Birds or animals may travel some distance from the point of impact before internal hemorrhage from ruptured organs results in death. It might be difficult to prove that someone intentionally ran down an animal (although - people have been prosecuted for hitting deer with cars for food!). These cases may be presented to determine cause of death when other causes are suspected. For example, animals are frequently shot from vehicles along roadways. Animals debilitated by toxins may die by the road as well. Also, as power lines frequently run along roads, electrocution would be another consideration for avian species.

- The necropsy findings with vehicle collisions are fairly predictable.

- On external examination, abrasions and hair loss may be evident from contact with the road surface.

- Victims usually have significant internal bleeding.

- Fractures may occur, but are not necessarily present.

- Automobile collisions most often produce injury by blunt trauma. Similar injuries are also caused by other means.

- Blunt trauma from instruments, such as bludgeoning, produces focal traumatic lesions, i.e. hemorrhage and fractures.

- Depending on the object used, the lesions may be too nonspecific to identify a specific type of weapon. However, instruments such as hammers and axes can leave distinctive wounds

ii. Electrocution

- Electrocution of raptors is very common. Birds found in association with power lines may or may not die from direct electrical insult resulting in thermo-electrical burns. On the other hand, some raptors may travel some distance after having been electrocuted.

- Electrocution in wildlife forensics most commonly involves power lines and avian species.

- Power lines often run parallel to roadways, thus the two causes of death must be differentiated.

Examination of the Wounds

- Power poles and lines are a common place for birds, especially raptors, to light, hunt and even nest.

- Birds are exposed in these locations and easily can be shot from their perches. Thus, gunshot is another rule-out in these cases.

- The mechanism of death in cases of electrocution is ventricular fibrillation. Inter-costal spasm and subsequent respiratory arrest also may be involved.

- Terminal hemorrhage at the base of the heart (reported in eagles) and muscle trauma may be observed due to violent contraction or subsequent falls. . Haemopericardium is a common finding in electrocuted eagles.

- Grossly, burns may or may not be present. Focal burns are accompanied by singeing and curling of feathers. As these birds may fall from great heights, secondary trauma may also be present.

- Burns, when present, may be as small as 1 to 2 mm in diameter or may envelope the entire carcass. Feathers around the primary lesion of contact are characteristically burnt and curled. Additionally, tissues along the tendons may appear "cooked". Subcutaneous tissues may be "pocketed" around seared fatty tissue.

iii. Cutting Injuries

- Tearing of skin must be differentiated from "cutting of the skin" which implies human intervention.

- Injuries caused by sharp-edged instruments are most commonly encountered when animals are incised along the neck as part of the field dressing procedure or to inflict a final fatal wound.

- Incision injuries also are produced when body parts are removed to harvest specific portions or to conceal an animal's species or sex.

- First, an investigator must determine whether an injury was inflicted while the animal was still alive or post-mortem. This information can be determined by examining the wound edges. In a living animal, the wound margins are infused with blood. In contrast, infusion does not occur following death.

- DNA analysis (forensic laboratories equipped to handle wildlife cases have PCR primers for many common species) will help in the identification of the head and body specimens when the head and body parts are found at different places.

iv. Trap-related Injuries

Snares and traps

- Trapping fur-bearing mammals is a legal means of wildlife harvest, provided that trappers abide by trap design regulations and take animals during the appropriate hunting season. Two general types of traps, the clamp-style traps and the snare-type mechanism are available. Clamp-style traps operate on a spring mechanism, whereby the jaws close in the animal. Most people are familiar with leg hold traps. Another type, called conibear trap, is designed to grasp animals around the body rather that an appendage. Many designs and styles exist for both types of traps. As the name implies, snares operate by creating a locking noose.

- Snares leave patterned lesions around the neck of trapped animals.

- Snares that are correctly set should be free running and should only catch the species for which they were designed. But in most of the cases snares are poorly placed, indiscriminate in what they trap and not be free running due to poor maintenance or deliberate use of illegal self-locking snares. When a non-target animal is found in a snare, the role of the forensic practitioner is to catalogue the injuries and to attempt to give an estimate of the duration of the entrapment.

- The cage traps may be legally used to take certain wild birds. They may sometimes be used with a live decoy bird, such as a crow, to entice other birds of the same species into the trap. Birds used as decoys or caught in such traps are at risk of death from starvation, dehydration or exposure if the traps are not run in accordance with the legal requirements and checked on a daily basis. Wild birds reduced into captivity in this manner are subject to welfare legislation and there have been convictions in relation to the mistreatment of decoy and trapped birds in cage traps.

- Daily checks should release non-target species, such as raptors. Where such species are suspected to have been left in such traps for extended periods consider the presence of faecal material and regurgitated pellets which may give an indication of the length of captivity. The physical condition of birds in a cage trap (particularly where no food is available), along with damage to feathers or around the beak due to repeated collisions with the cage wire, may give an indication that a bird has been present for an extended period. Occasionally, there is deliberate injury to decoy birds, such as fractured wings, and the forensic practitioner may need to establish whether such fractures occurred before death.

Examination of the Wounds

- On external examination, such lesions may not be evident on gross examination of the body of the animal. Removal of the skin is necessary to visualize abnormalities. On the other hand, some leg hold trap lesions may be quite gruesome, resulting in open fractures and self-mutilation.

v. Involvement of scavenger or predator

The carcass of animals allegedly killed by predators should be carefully examined for lesions which indicate premortem hemorrhage. Scavenger damage must be differentiated from actual predator attack by assessing the pattern of attack, the presence of a lethal wound caused by the suspect predator, and the demonstration of hemorrhage associated with the wounds indicating the animal was alive when attacked.

Determination of an animal's involvement, or lack thereof, in another animal's or a human's death may be of great importance. It may be necessary for investigators to determine if a predator was responsible for the killing or if a predator was scavenging a carcass that died from some other etiology. Also, it may be necessary to determine what kind of animal(s) or even what individual was responsible. Examination of pattern of attack, bite marks, hair samples, and saliva DNA will help the investigators to identify the culprit.

The pattern of attack can be inferred from the pattern of the bite wounds. The wounds consist of a large area over the thoraco-lumbar spine and a second area over the neck. Two sets of large puncture wounds are identified over the affected areas. At the time of necropsy, samples are taken in the region of the bite wound for possible saliva DNA extraction. Hunting behavior varies between predator species. Ex. Canids (wolves, domestic dogs, etc.). slow their prey by attacking the distal limbs. This may be evident at necropsy by damage to the region of the gastrocnemius muscles as is seen often in predated artio-dactylids. The lesions may be most apparent in the tissues underlying the skin, with little external evidence of trauma (another feature of the bite wounds of canids). Canids may hunt in groups while most cat species are solitary hunters, African lions being one exception. Felids, such as mountain lions, frequently drag down larger prey by lunging onto the back. The final kill is made by a bite to the cervical region.

The predators can also be identified by examining the hair. Many keys have been produced to help investigators identify the source of animal hair samples.(hair samples may be available from the scene or on the victim).

vi. Dog bite injuries

Documentation of bite-- fatal bite wounds must include skinning of the carcass because much of the damage is done without skin puncture.

In general, dogs' teeth are relatively blunt. As the dog seizes its prey, the skin stretches under the pressure of the teeth allowing extensive damage to muscles, bones and internal organs without necessarily puncturing the skin. Consequently, in such cases, when the outside of the animal is examined there may be little or no evidence of skin holes, bleeding or other external injury. Hares, deer, sheep, badgers, foxes and other dogs may all be subject to attacks by dogs. More than one dog may well be involved in an incident. The type of dog involved also has a significant bearing on the type of injuries inflicted. A fox attacked by a lurcher or foxhounds above ground will have very different injuries to the predominantly facial and front end injuries received when faced with a terrier underground.

A postmortem examination may be essential to confirm or refute any allegations. This can provide detailed evidence in relation to the injuries and their potential cause. Estimates of the size of the mouth of the biting animal may be calculated by measuring the inter-canine distance of 'bite pairs'. With live animals suspected of being used to attack dogs or other animals, a detailed examination of injuries may provide vital evidence. Dogs repeatedly used for fighting may suffer characteristic and often severe facial injuries. DNA testing may be used to match saliva samples left on attacked animals back to individual dogs. Similarly the tissue or hair found between the teeth of a dog suspected of being used for fighting or attacking other animals may provide very important evidence. Such samples can potentially be identified by morphological examination or DNA testing, and in some cases linked back to individual animals held by a suspect.

6 | Types of Firearms and Terminology

When a forensic investigation involves a shooting, ballistics becomes an important facet of the investigation. Forensic ballistics is the science dealing with fire arms. Ballistics is a term that refers to the science of the flight path of a bullet. The flight path includes the movement of the bullet down the barrel of the firearm following detonation and its path through both the air and the target. Knowledge of the different types of ammunition available is critical to assessing such cases.

In order to properly assess gunshot wounds, it is necessary to have an understanding of the weapons and ammunition that are commonly used. Modern firearms are manufactured in a variety of shapes and sizes to fit multiple purposes. This information will greatly assist an investigator with analyzing the characteristics of inflicted wounds and collecting relevant information and evidence at the time of necropsy. Tracing the path of a bullet is important in a forensic examination. It can reveal from what direction the bullet was fired, which can be vital in corroborating the course of events in the crime or accident. Recovery of bullets can be a very useful part of forensic ballistics. By collecting the appropriate evidence during necropsy, the investigator potentially can match a weapon to the accused, prove that a fatal wound was inflicted by a specific weapon or type of weapon, and prove off-season weapon violations. A variety of bullet designs exist, some that are specific to the firearm. Furthermore, the scouring of a bullet's surface as it encounters the grooves of the firearm barrel can produce a distinctive pattern that enables a bullet to be matched with the firearm. A weapon recovered from a suspect can be test fired and the bullet pattern compared with a bullet recovered from the scene to either implicate or dismiss involvement of the firearm in the crime. The distance that a bullet can travel depends on its speed. A higher speed imparts more energy to the bullet. The frictional resistance of the air and the downward pull of gravity will take longer to slow the bullet's flight, as compared to a bullet moving at a lower initial velocity.

Generally, a bullet fired from a rifle will carry more energy than a bullet fired from a handgun. Expansion of the exploding gunpowder generates pressure, which is measured as the force of the explosion that pushes on the area of the bullet's base. This area is essentially the diameter of the barrel of the firearm, which remains constant. A longer barrel will produce a faster moving bullet.

Once a bullet leaves the rifle or gun barrel, the aforementioned frictional and gravitational forces begin to slow its speed, producing a downward arc of flight. The frictional force is affected by the bullet's shape. A blunt shape will present more surface area to the air than will a very pointed bullet.

Another factor that affects the flight of a bullet is called yaw. As in an orbiting spacecraft or a football tossed through the air, yaw causes a bullet to turn sideways or tumble in flight. This behavior is decreased when the object spins as it moves forward (the spiraling motion of a football). The barrel of a rifle or gun contains grooves that cause the bullet to spin. More damage results from a bullet that is tumbling rather than moving in a tight spiral

The shape of a typical bullet—much like a football with one end blunt instead of tapered—is a compromise that reduces air resistance while still retaining the explosive energy that allows the bullet to damage the target.

The composition of a bullet is also important. Lead is commonly used to form the core of bullets. However, because it tends to deform, the blending in of other metals (typically antimony and copper) produces a bullet that can withstand the pressure of flight and impart high energy to the target upon impact .

Copper is often used to jacket the inner lead core of a bullet. However, some bullets are deliberately made without this full metal jacket. Instead, the bullet has

Types of Firearms and Terminology

a tip made of lead or a tip that is hollow or very blunt. These bullets deform and break apart on impact, producing more damage to the target than is produced by a single piece of metal. This is because the bullet's energy is dissipated within a very short distance in the tissue.

Forensic and medical examiners are able to assess the nature of tissue damage in a victim and gain an understanding of the nature of the bullet used.

Identifying the manufacturer and calibre of a submitted bullet is an important examination conducted in a forensic laboratory by Firearm and Tool mark Examiners.

Firearms
Handgun (A firearm designed to be fired from the hand). There are handguns made specifically for target competition or hunting, These are compact for concealability and ease of carrying. Cases of handgun wounds are more common in human crime; however, cases involving animals may be presented, especially when a handgun is used to inflict the fatal or "final" wound. The two most common defensive handguns are the double action revolver and the semiautomatic pistol.

Pistol: Synonym for a handgun that does not have a revolving cylinder. Pistol bullets are propelled at 1000 feet per second	**Revolver:** Handgun that has a cylinder with holes to contain the cartridges. The cylinder revolves to bring the cartridge into position to be fired. This is "single-action" when the hammer must be cocked before the trigger can fire the weapon. It is "double-action" when pulling the trigger both cocks and fires the gun.

Semiautomatic pistol	Barrel length is smaller for concealability and longer for accuracy or energy. The ejector rod under the barrel is used to eject fired cartridges before reloading. A revolver may weigh less than 1 lb to more than 4 lbs. The cylinder contains five or six holes for the cartridges and can be swung out for easy reloading. This must be a conscious act, so that no empty cartridge cases will be found at a crime scene unless the assailant stopped to reload. The revolver bullets are propelled at a velocity of 500 feet per second
The advantage of semiautomatics is the use of recoil generated by the fired cartridge to eject the empty cartridge case, load the next cartridge, and cock the hammer. This is more conducive to firing multiple shots, so many are designed to carry 15 to 19 rounds.	
The handle, or butt, is more important here because it contains the magazine holding the cartridges. Safety mechanisms prevent accidental firing. Some lock the hammer, while other designs lock the trigger.	There is a gap between cylinder and barrel to allow the cylinder to turn freely, but this also allows gases to escape laterally, which at close range may deposit gunshot residue on surrounding structures and allow the forensic pathologist to reconstruct the scene.

Even on open ground ejected cases may be difficult to find, as they typically roll into a hiding place such as grass or small depressions in the ground. Thus, ejected cases will virtually always be left behind at the scene, but must be searched for diligently.

Long guns

A general term that includes bot rifles and shotguns. These comprise the majority of weapons in wildlife forensic cases. *Most modern pistols, revolvers, rifles and some shotgun barrels have what are called *rifling* in their barrels.*

Smooth bore weapons- Shotguns	Rifled weapons
Smooth bore weapon has a smooth bore (there is no rifling in its barrel).Ex. shotgun A shot gun is a smooth bore weapon. Shot guns are of different types and may be 1. short or long barreled, 2. Double or single barreled, 3. Full choked, half choked or non chocked and 4. Muzzle loading or breech loading types. A shot gun cartridge is made of a special type of paper and contains lead shots. The propellant material is called black powder which consists of charcoal (15%), potassium nitrate(75%) and sulphur(10%). It burns with production of a lot of smoke. Some powder is only partly burnt or not burnt at all. Hence it is called black powder. Company made shot gun cartridges use lead, while home made ones employ glass beads or stones. If cartridges are home made, probably the gun is also home made, crude and indigenous one.	A rifled weapon is so called because of what is called rifling present in its barrel. Rifles vary from one make of weapons to another in number, direction, depth and width of rifling Rifling is made up of lands and grooves. The grooves are spiral, are cut in the inside of the barrel, leaving elevated portions between the called 'lands'. These lands bite into the bullet as it is propelled through the barrel. Rifling may be clock wise or anticlockwise. Rifling gives the bullet, a spin, greater power of penetration, a straight trajectory and prevents it from wobbling as it travels in the air (it gives the bullet gyroscopic steadiness). As the bullet is propelled through the barrel, the surface of the bullet becomes marked with grooves corresponding in number and direction with the lands in the barrel. These indentations which are transferred to the bullet from the barrel are peculiar to that weapon alone and to no other and this is therefore called the "thumb print" of the weapon. The rifled weapons are pistol, revolver and rifle. The rifle bullets are propelled at 2500 feet per second.

Smooth bore weapons- Shotguns	Rifled weapons
The calibre of a shot gun is the size of the lead balls which will exactly fit into the barrel and by the number of such balls that can be made out of one pound of lead. Ex. A 12-bore gun is one the calibre of which is that of a lead ball of such size that 12 such balls can be made out of 1 pound of lead. The greater this number, the less the caliber of the weapon Ex. A 20 bore gun is of smaller caliber than a 12 bore gun. **Gauge** Shotguns are classified by an entirely different measure known as gauge. Gauge is measured by the number of lead balls with the same diameter as the barrel of the weapon that would be required to equal one pound (refers to the diameter of the barrel on a shotgun in terms of the number of lead balls the size of the bore it would take to weigh one pound (10 gauge, 12 gauge, etc.) The most commonly encountered weapons are the 20, 16, 12, and 10 gauge shotguns. The larger the gauge, the smaller the bore of the weapon. The exception to this is the 410 shotgun, a smaller weapon that is a direct bore measure.	Calibre or bore or inner diameter of rifled weapon is usually mentioned in terms of an inch. The calibre is measured between lands and not grooves. The size of a weapon is measured by the size of its bullets in hundredths of an inch (English measure). This system is used to classify most firearms, with the exception of shotguns. For example, a 38 calibre weapon fires a bullet that is 0.38 inches in diameter. The European system, uses metric measurements. A weapon with a European measure that is familiar to most people is the 9 millimeter handgun. Rifles vary from one make of weapon to another in number, direction, depth and width. Grooves are cut into the barrels of rifles and some shotguns to place a spiral motion on a projectile as it leaves the barrel. This promotes a more linear flight path. Rifling marks left on a projectile can help in matching a projectile to a specific weapon.

Types of Firearms and Terminology

Smooth bore weapons- Shotguns	Rifled weapons
As with rifles, shotguns are constructed with the same basic four components. Shotguns are capable of firing pellets, larger metal balls (shot), or slugs. These projectiles are loaded into shells that contain propellant (gunpowder) and other components required for discharge. Shotguns have a similar external appearance to rifles, but differ in the lack of rifling inside the barrel, which is the basis for their legal definition. A shotgun shell may contain one large projectile (called a slug), a few pellets of large shot, or many tiny pellets. Shotguns are available in single shot (break action), double barrel, pump action, and semiautomatic. A gun with a smooth bore that shoots cartridges that contain "shot" or small metal pellets (of lead or steel) as the projectiles.	The basic components of modern rifles include the barrel, magazine, loading and ejecting apparatus, and stock. The stock is rested against the shoulder to bear the force of the recoil when the weapon is discharged. The magazine is the location where ammunition is stored and fed into the loading and ejection port. Several different loading and ejection mechanisms exist, including bolt action, lever action, automatic loading. Modern rifles fire bullets that are loaded into cartridges.

High velocity firearm	Low velocity firearm
These weapons include most of the popular hunting rifles and have a muzzle velocity of greater than 2,200 feet per second	Shotguns and muzzle loader weapons fire heavier, slower projectiles with a muzzle velocity of around 1,300 feet per second or less.

Muzzle velocity - The speed at which a projectile leaves the barrel of a weapon. Muzzle speed usually is given in feet per second.

Muzzle loader – These are old-fashioned single shot firearms that are loaded through the distal end of the barrel. With traditional muzzle loaders, each component, gunpowder, wadding, projectile, patch, is loaded separately and placed down the barrel by a ram rod (seen mounted below the barrel). The process must be repeated with each single shot. Modern muzzle loaders are available that lend to quick loading and impressive accuracy.

Muzzle loader

Cartridge – Also called a "round". This is a self-contained unit that is made up of an outer casing, projectile (bullet), propellant (gunpowder), and primer (source of ignition). Only the projectile is fired from the gun. The remainder of the spent cartridge is ejected from the weapon.

Cartridge casing – This component (outer shell or covering) of the cartridge is usually composed of brass and can be vital in linking a weapon to a specific case. This helps to keep the various components of the cartridge in place and also provides a water proof cover for the contents. The cartridge ejection mechanisms of rifles leave characteristic marks that are analogous to the human fingerprint. The cartridge case of a rifled weapon is made of an alloy of copper and zinc.

Cartridges are usually given a name or cartridge designation by their developer, who is more often than not the manufacturer of a firearm. The cartridge designation typically includes a numerical value to indicate the approximate diameter of the bullet and will often include the manufacturer's name.

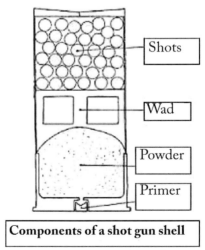

Components of a shot gun shell

Jacket – The metal covering over a bullet that overlies a core of different consistency. Jackets are used to manipulate the degree of deformation that occurs upon impact.

Types of Firearms and Terminology

Shotgun pellets – Shotgun pellets vary in size from 1.27 mm bird shot to 9.14 mm buckshot. The large diameter pellets are used to hunt larger game such as deer. Shotgun pellets may be made from lead, steel, and various alloys. As mentioned above, lead birdshot has been declared illegal in waterfowl hunting.

Shot cup – A shot cup is a plastic sleeve that is loaded into a shotgun shell prior to the projectiles. The cup holds the projectiles and facilitates their uniform discharge from the barrel. The shot cup falls away when a shell is discharged. Cup sizes are specific for different gauge shells.

Wadding – Wadding is the material (compressed paper) placed between the propellant (powder) and the projectile (shot). It is used in shotgun shells and muzzle loader-type firearms to distribute force on the projectile. Nowadays the wadding and shot cup usually is formed into a single unit. Shot string – As pellets are discharged from a shotgun, they spread out over distance. The area covered by the pellets varies according to distance from the target and what is referred to as the choke. The pattern of the pellets as they reach a target is referred to a shot string. An animal's location within the shot string will determine the number of pellets.

Standard handgun and rifle cartridges

The metal casing encloses the powder, above which the bullet is seated. The powder is ignited through the flash hole when the primer is struck. A case with a rim is found with revolver and lever action rifle cartridges, and also with some bolt action and semi-automatic rifles.

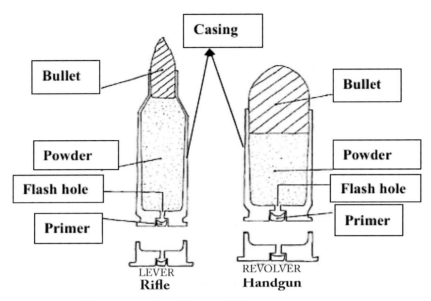

Components of a cartridge

Both smooth bore and rifled weapons employ cartridges, but of different types. The outer shell or covering of the cartridge is called the cartridge case. This helps to keep the various components of the cartridge in place and also provides waterproof cover for the contents

Shot gun cartridge	Rifled weapon Cartridge
A shot gun cartridge is made up of a special type of paper. It contains lead shots. The propellant material is called the black powder which consists of charcoal (15%), potassium nitrate (75%) and sulphur (10%). It burns with production of lot of smoke. Some of the powder is only burnt or not burnt at all. Hence it is called black powder. In between the shots and the powder is a wad made up of compressed paper or felt. This serves the purpose of separating the shots from the powder.	The cartridge case of a rifled weapon is made of an alloy of copper and zinc. The cartridge has a bullet which is made of lead and has a coating of a cupro-nickel alloy. It is conical in shape. The propellant material known as smokeless powder is nitrocellulose. This burns almost completely with production of much less smoke than black powder. Consequently the injuries caused by these weapons do not cause much smudging and tattooing.

After detonation of a cartridge the bullet or shots are propelled through the barrel of the weapon by the expanding gases generated by the explosion of gun powder. These gases escape out of the muzzle immediately after the bullet.

Bullet

The projectile that is loaded into a cartridge and fired from a firearm. Bullets are available in various shapes, e.g., hollow point, and weights (measured in grains) Bullets / projectile. are shaped or composed differently for a variety of purposes.

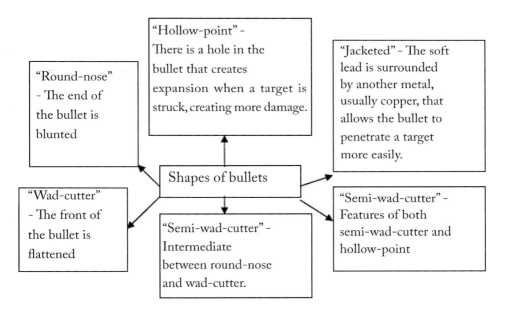

The bullets are made of lead and a coating of a cupronickel alloy. The propellant material is a smokeless powder (nitrocellulose). This burns almost completely with production of much less smoke than black powder.

Shapes of bullet

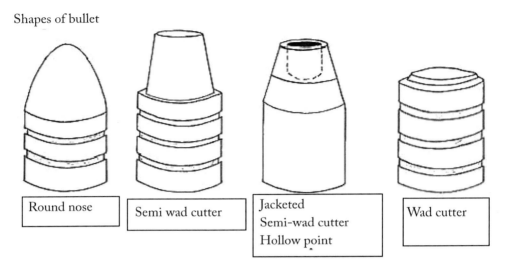

When a bullet is submitted for comparison to a firearm, one of the first examinations conducted will be to determine the bullet's caliber.

Bullet design is important in wounding potential. The distance of the target from the muzzle plays a large role in wounding capacity, for most bullets fired from handguns have lost significant kinetic energy (KE) at 100 yards, while high-velocity military .308 rounds still have considerable KE even at 500 yards. Military and hunting rifles are designed to deliver bullets with more KE at a greater distance than are handguns and shotguns.

The type of tissue affects wounding potential, as well as the depth of penetration. Specific gravity (density) and elasticity are the major tissue factors. The higher the specific gravity, the greater is the damage. The greater the elasticity, the less is the damage. Thus, lung tissue of low density and high elasticity is damaged less than muscle with higher density but some elasticity. Liver, spleen, and brain have no elasticity and are easily injured, as is adipose tissue. Fluid-filled organs (bladder, heart, great vessels, and bowel) can burst because of pressure waves generated. A bullet striking bone may cause fragmentation of bone and/or bullet, with numerous secondary missiles formed, each producing additional wounding.

The speed at which a projectile must travel to penetrate skin is 163 fps and to break bone is 213 fps. To penetrate the thick hide and tough bone of an elephant, the bullet must be pointed, of small diameter, and durable enough to resist disintegration

Types of Firearms and Terminology

However, such a bullet would penetrate most human tissues like a spear, doing little more damage than a knife wound.

Round nose bullets provide the least braking, are usually jacketed, and are useful mostly in low velocity handguns. The wad-cutter design provides the most braking from shape alone, is not jacketed, and is used in low velocity handguns (often for target practice). A semi-wad-cutter design is intermediate between the round nose and wad-cutter and is useful at medium velocity. Hollow-point bullet design facilitates turning the bullet "inside out" and flattening the front, referred to as "expansion." Expansion reliably occurs only at velocities exceeding 1200 fps, so is suited only to the highest velocity handguns.

Wounding is an extremely complex situation with variables of bullet size, velocity, shape, spin, distance from muzzle to target, and nature of tissue. These factors are interrelated, and the wounding potential may be difficult to predict even under controlled test conditions. In an actual forensic case, few of the variables may be known, and it is up to the medical examiner to determine what can be known from examination of the evidence.

Handgun Ballistics

These weapons are easily concealed but hard to aim accurately, especially in crime scenes. Most handgun shooting occur at less than 7 yards, but even so, most bullets miss their intended target. The two major variables in handgun ballistics are diameter of the bullet and volume of gunpowder in the cartridge case.

Shotgun Ballistics

Wounding is a function of the type of shot, or pellets, used in the shotgun shell. Weight, in general, is a constant for a shell. The spread of the pellets as they leave the muzzle is determined by the "choke" or constriction of the barrel at the muzzle (from 0.003 to 0.04 inches). More choke means less spread. Full choke gives a 15 inch spread at 20 yards, while no choke gives a 30 inch spread at the same distance. A "sawed-off" shotgun has a very short barrel so that, not only can it be concealed more easily, but also it can spray the pellets out over a wide area, because there is no choke.

Close range-less than 4 feet	Intermediate range 4-12 feet	Beyond 12 feet
The pellets essentially act as one mass. Entrance wound would be about 1 inch diameter, and the wound cavity would contain wadding.	The entrance wound is up to 2 inches diameter, but the borders may show individual pellet markings. Wadding may be found near the surface of the wound.	Choke, barrel length, and pellet size determine the wounding.

Pellets, being spherical, are poor projectiles, and most small pellets will not penetrate skin after 80 yards.

The spread may carry them away from the target. Thus, close range wounds are severe, but at even relatively short distances, wounding may be minimal. Range is the most important factor. A rifled slug fired from a shotgun may have a range of 800 yards.

Shotgun slugs can produce significant injury, because of the slug's size and mass. At close range, survival is rare. In treating shotgun injuries, it is necessary to remember that the plastic shell carrier and the wadding (which may not appear on radiographs) can also cause tissue damage and may need to be found and removed.

Rifle Ballistics

Many different cartridges are available using different loads and bullet designs. The wounding potential of projectiles is a complex matter. Bullets fired from a rifle will have more energy than similar bullets fired from a handgun. Bullet travel through a gun barrel is characterized by increasing acceleration as the expanding gases push on it. Up to a point, the longer the barrel, the greater is the acceleration. Thus, cartridges designed for hunting big game animals use very large bullets.

Terminal ballistics (hitting the target)

Bullet velocity and mass will affect the nature of wounding. Velocity is classified as low (<1000 fps), medium (1000 to 2000 fps), and high (>2000 fps). An M-16 rifle (.223 cal) is designed to produce large surface wounds with high velocity. Low mass

Types of Firearms and Terminology

bullets that tumble, cavitate, and release energy quickly. A hunting rifle (.308 cal or greater) would have a larger mass bullet to penetrate a greater depth to kill a large game animal at a distance.

Definitions of the Terms

Action - The part of a firearm that loads, fires, and ejects a cartridge. Includes lever action, pump action, bolt action, and semi-automatic. The first three are found in weapons that fire a single shot. Firearms that can shoot multiple rounds ("repeaters") include all these types of actions, but only the semi-automatic does not require manual operation between rounds. A truly "automatic" action is found on a machine gun.

Wound ballistics which concerns the pathology of projectile impact in the tissues of the animal or is the study of the pathological effects of a projectile passing through a body. The basic types of guns used to kill animals are Rifles (low velocity and high velocity), pistols, air guns, and shot guns. Significant differences exist in the wounds caused by high velocity center fire rifle bullets and low velocity bullet wounds from the common .22 rim fire bullets, black powder rifle bullets, most pistols' bullets and shotgun slugs. Shotgun pellets may be multiple but are generally of a relatively low velocity.

Barrel - The metal tube through which the bullet is fired.

Black Powder - The old form of gunpowder invented over a thousand years ago and consisting of nitrate, charcoal, and sulfur.

Bore - The inside of the barrel. "Smooth bore" weapons (typically shotguns) have no rifling. Most handguns and rifles have "rifling".

Breech - The end of the barrel attached to the action.

Butt or butt stock - The portion of the gun which is held or shouldered.

Centre fire - The cartridge contains the primer in the center of the base, where it can be struck by the firing pin of the action.

Chamber - The portion of the "action" that holds the cartridge ready for firing.

Choke - This term refers to the constriction at the end of the shotgun barrel to prevent dispersion of shots. The extent of the choke affects the spread of the pellets following discharge from the barrel, thus manipulating the shot string.

Double-action - Pulling the trigger both cocks the hammer and fires the gun.

Double barrel - Two barrels side by side or one on top of the other, usually on a shotgun.

Hammer - A metal rod or plate that strikes the cartridge primer to detonate the powder.

Ignition - The way in which powder is ignited. Old muzzle-loading weapons used flintlock or percussion caps. Modern guns use "primers" that are "rim fire" or "center fire"

Lands and grooves - Lands are the metal inside the barrel left after the spiral grooves are cut to produce the rifling.

Magazine - This is a device for storing cartridges in a repeating firearm for loading into the chamber. Also referred to as a "clip"

Magnum - An improved version of a standard cartridge which uses the same calibre and bullet, but has more powder, giving the fired bullet more energy. Magnum shotgun loads, however, refer to an increased amount of shot pellets in the shell.

Muzzle - The end of the barrel out of which the bullet comes.

Powder - Modern gun cartridges use "smokeless" powder that is relatively stable, of uniform quality, and leaves little residue when ignited. For centuries, "black powder" was used and was quite volatile (ignited at low temperature or shock), was composed of irregularly sized grains, and left a heavy residue after ignition, requiring frequent cleaning of bore.

Primer - A volatile substance that ignites when struck to detonate the powder in a cartridge. "Rim fire" cartridges have primer inside the base, while "center fire" cartridges have primer in a hole in the middle of the base of the cartridge case.

Rim fire - The cartridge has the primer distributed around the periphery of the base

Safety - A mechanism on an action to prevent firing of the gun..

Sights - The device(s) on top of a barrel that allow the gun to be aimed.

Types of Firearms and Terminology

Silencer - A device that fits over the muzzle of the barrel to muffle the sound of a gunshot. Most work by baffling the escape of gases.

Single - action - The hammer must be manually cocked before the trigger can be pulled to fire the gun.

Stock - A wood, metal, or plastic frame that holds the barrel and action and allows the gun to be held firmly.

Slug - Slugs are another type of projectile that may be loaded into shotgun shells. There are two basic types, the Foster (right) and Sabot slugs (left). Foster slugs are grooved so that they rifle when fired through a smooth-bore shotgun. Sabot slugs are discharged with a polyurethane sleeve. Slugs primarily are used to hunt bigger game.

Smokeless powder - Refers to modern gunpowder, which is really not "powder" but flakes of nitrocellulose and other substances. Not really "smokeless" but much less so than black powder .

Trajectory - is the direction of the projectile relative to the shooter and victim. Establishing the trajectory is very important in so called "self defense" cases. It may also be important in locating a bullet which has passed through and animal and is lost into the environment. Bullets, if found, are important evidence items which may be individually matched back to a suspect's firearm. Bullet entrance wounds have different characteristics than bullet exit wounds and the pathologist must differentiate entrance from exit wounds.

7 | Poisoning

Poisoning of protected wildlife is common and is a major wildlife forensics issue. Most (82%) of wildlife poisonings involve Organophosphorus (OP) or carbonate pesticides which are used to control livestock "predators" such as eagles, coyotes, bears, wolves and cougars. Field investigation is essential for the proper documentation of suspected poisoning cases. Information related to species affected, food habits of the species, clinical signs or death stance, post mortem condition of most of victims, scavenger evidence, legal pesticide use history in the area are all important to establish in order to narrow the list of possible toxic substances and the sources of such substances. A combination of field work, veterinary examination and chemical analysis is used to determine the underlying cause of death.

Poisoning may be caused by chemical agents or plant poisons. In such cases the diagnosis is given based on the gross lesions and one should not determine the presence or absence of poison, which will be done by the forensic laboratory (these are the only authorized places where materials of dead animals or human are examined to detect the poisons / chemicals).

The forensic investigator may encounter cases that range from intentional poisoning of pest species and secondary exposure of non-target species, to accidental exposure of wildlife by labeled use of pesticides, to cases involving massive environmental pollution. Investigators must have an idea of likely suspect toxins based on history, previous cases for a given region, and post-mortem examination.

Toxins most commonly used include carbamates, organophosphates (OP's), strychnine, compound 1080 (sodium fluoroacetate), anticoagulant rodenticides, thallium, and cyanide. Carbamates and OP's comprise the greatest percentage of toxicosis cases. Many of these chemicals are or have been used in agriculture.

Several factors may clue investigators into a suspicion of toxicosis. For example, a dead animal or localized group of dead animals in an apparently good nutritional state is highly suspicious of a toxin, especially if multiple species are involved. In some cases, investigators may locate a poisoned bait or have a good indication of agricultural pesticides commonly used in a given area.

Whenever a toxin is suspected, necropsy staff, as well as field personnel, should take precautions to prevent accidental human exposure. Samples should be taken in appropriate containers (acid-rinsed glassware) and toxin analyses should be conducted as soon as possible. .

When animals are directly affected by the application of a substance to poisoned bait, this is referred to as primary toxicosis. Secondary toxicosis occurs when scavengers feed on other animals affected by a original poison source.

A combination of field work, veterinary examination and chemical analysis is used to determine the underlying cause of death. At the scene of suspected wildlife poisoning make careful assessment of the following.

1. Any signs to suggest a poisoning incident like victim lying next to potential bait with signs of food in the mouth / crop.

2. Presence of dead flies or other insects on, under or close to a suspected bait (indicates the involvement of insecticide).

3. Any potential forensic evidence at the scene like tyre, foot wear, or spade impressions, discard items etc. Take photographs and proper sketch plan.

Points to Remember

- Most acute poisons leave little or no visible pathological changes. Therefore the objectives of the pathological evaluation is to collect appropriate samples for chemical analysis and identification of the toxin, document and identify the possible source of the poison, and determine if the victim is a primary or secondary target (intent of suspect).

- Identify the source of poison (ingestion and contact),

- Detect lethal poison in illegally killed animals and

- Determine if primary or secondary target. The primary is by consumption of poison bait and the secondary is due to consumption of another animal which was primary target of poison.

- Crop and/or stomach content is the most frequently tested material for ingested poisons. Carbamate and organophosphate pesticides are often absorbed directly through the crop lining in birds. When the crop and stomach are empty of food items, the tissues lining the crop or stomach may be analyzed for the presence of pesticides.

- If pesticides are present in the lining, a water source might be considered along with the possibility of vomiting of the original content.

- Skin of the feet of birds may be tested for absorbed pesticide residue (cutaneous absorption). The cutaneous tissues of the feet of raptors may be the only tissues that retain traces of organophosphate and carbamate pesticides in decomposed or scavenged carcasses.

- Tissue levels of organic pesticides may vary significantly post mortem due to decomposition of the molecular structure of the chemical.

- Brain tissue may be collected for cholinesterase determination.

- Identification of items such as hair, tissue, feathers, seeds, etc. and their location within a victim's digestive track is an important forensic consideration. These items, as well as the suspect material for chemical analysis, become additional evidence items which may be passed on to other analysts.

- Documentation of presence, collection, position in the digestive track, etc. and proper establishment of a chain of custody are important.

- Since the defense also has a right to evaluate the evidence if they choose, duplicate samples of material which is consumed in analysis should be collected on a routine basis.

- Carbofuran and aldicarb are the most commonly detected pesticides in cases of intentional wildlife poisonings. Famphur, an unrestricted organophosphate pesticide used widely in the cattle industry, is also used to illegally poison pest birds which often results in secondary poisoning of predatory or scavenger birds. Strychnine, avitrol and anticoagulant rodenticides are pesticides that tend to be seen in secondary poisoning cases of protected wildlife where they are used legally to control pest rodents or birds. Accidental poisoning of wildlife occurs associated with legal agricultural use of pesticides.

- Carbofuran is available in both granular and liquid forms. It is often poured on carcasses or incorporated into baits to kill scavengers such as coyotes, foxes, raccoons, eagles, vultures, etc. Aldicarb (seen as small black granules), is frequently observed in baits. The granular form is resistant to biological degradation and may remain in the environment for extended periods, especially in arid climates. Both of these compounds may kill birds within minutes of ingestion leading to the "circle of death" surrounding a baited carcass.

- Strychnine is frequently observed in victims as green or blue green round grain seeds which are commercial preparations intended for rodent control. Nuisance waterfowl often are intentionally poisoned with strychnine impregnated grain. Raptors feeding on rodents which have fed on the seeds are victims of secondary poisoning.

- Eagles and other scavengers are particularly susceptible to pentobarbital poisoning. Eagles feeding on improperly disposed large animal carcasses after euthanasia die in large numbers. (due to improper directions to the client by veterinarians who are found liable for eagle kills).

- Oil contamination of birds after oil spills and due to entrapment in oil tanks, separation ponds or other facilities associated with petroleum production is a major need of forensic documentation. Proper documentation of cause or causes of death in such cases must consider the possibility of death from other causes and secondary contamination of the carcass post-mortem. Numerous forms of oil or petroleum products may be spilled into the environment with variable toxicities or physical effects. The cause of death in oil cases are hypo or hyperthermia, exhaustion, inhalation (volatile and non-volatile) ingestion (short term or long term toxic effects) and combination of all these.

- A full postmortem examination will be conducted in most suspected poisoning cases before submitting samples for toxicology. The postmortem examination will ensure that other evidence such as trauma or natural disease is not overlooked. The pathologist can collect and submit the appropriate samples to the toxicology laboratory. Radiographs may indicate a bird has recently eaten and may cause suspicion that it has been poisoned.

Where shot animals are used to prepare poisoned baits, then radiographs showing shotgun pellets in the gullet or gizzard may be indicative of illegal poisoning incidents.

Materials to be sent

The stomach and its contents after tying both the ends, 30 cm of Ileum and colon and their contents with ends tied
Rumen contents – 2 Kg in ruminants
0.5 Kg of liver
One or both the kidneys
If plant poisoning is suspected, either dried (under shade) or fresh plants
Urine with equal quantity of spirit or if the spirit is contraindicated 4 g of thymol may be added and finally faeces in spirit.
Lungs and blood from the heart
Skin and subcutaneous tissue if poison is injected
Long bones in sub acute and chronic arsenic and antimony poisoning
Hair in sub acute and chronic mineral poisoning

In case of strychnine / nox vomica poisoning - Heart and a portion of brain

In Carbon monoxide, coal gas, Hydrocyanic acid, alcohol or chloroform poisoning, materials (lungs and blood from the heart) are to be sent immediately and without any preservatives. In case of alcohol poisoning, cerebrospinal fluid has to be sent.

In case of suspected alcohol, phosphorus, paraldehyde, acetic acid or phenol and other drugs of phenol group poisoning, a saturated solution of sodium chloride is used as preservative.

Poisoning 263

In criminal abortions, sticks and other foreign bodies recovered from the genital tract should be collected and preserved in separate bottles after drying.

Note: The viscera collected should be kept safely (if not sent to the chemical examiner) for 6 months and then destroyed after obtaining the written order from the magistrate.

The notes written while conducting the post mortem examination should be retained and preserved especially that of vetero-legal cases since these will help later when the evidence is given in the court. From the notes, the postmortem report should be prepared and given as early as possible. It should not be delayed. It should be mentioned in the report that the final opinion will be given based on the results of the report of the chemical examiner / laboratory. Always avoid over writing in the report. The report should be prepared in duplicate or triplicate.

Detection of toxic chemicals is a difficult task and adequate specimens are necessary. The samples required depends on the type of toxin or poison suspected and whether the animal is alive or dead. The veterinarian should be quite clear as to what sample will be needed and state clearly in his request form the estimation required. One has to submit the suspected source of the poison. Any tissue submitted should be at least 50 g in weight and in a clean container. In any poisoning, serum- 5ml, whole blood- 10 ml (heparinised), and urine 50 ml, frozen tissues or tissues preserved in alcohol or formalin, feed and water sample(unpreserved) and dry dung are sent.

Materials to be sent to the forensic laboratory in various poisoning cases

Poison	Materials to be sent
Arsenic	Acute: Kidney, liver (1Kg), vomit and stomach contents (1 kg). In chronic poisoning and if animal is alive, hair sample, urine, skin and long bone.
Copper	Liver tissue, kidney, whole blood
Strychnine	All stomach contents, liver, kidney and all urine under refrigeration
Alkaloids	Liver, stomach contents, brain

Poison	Materials to be sent
Cyanide	Stomach or rumen contents / intestinal contents, portion of liver and muscle. Stomach contents are not conclusive. Cyanide is lost rapidly after death especially from liver. Hence all samples should be presented in a solution of 1% mercuric chloride before being dispatched or submit fresh tissues. Whole blood 10 ml and Forage / silage 100 g.
Heavy metals (mercury, lead, bismuth)	50 – 100 g of liver. Kidney, stomach contents under refrigeration. Avoid contamination with stomach contents. Send stomach contents in a sealed container separately. Lead: If animal is alive, 10 ml heparinised whole blood. Do not use citrate. If dead, send kidney and liver. Examine rumen / intestinal contents and anything suspicious, collect and send them to laboratory. At least 20 g of tissue is required and send in plastic container. Mercury: Fresh liver, kidney, stomach and intestinal contents and feed.
Insecticides	Fatty tissue, liver, brain stomach contents and heparinised blood samples. Use chemically clean glassware. Avoid plastic containers
Hydrocyanic acid	Plant material, stomach / rumen contents, blood and liver-fresh material to be submitted. Specimen should be frozen.
Nitrates / Nitrites	Stomach contents, kidney, 10 ml clotted blood / whole blood and material containing nitrates- forage / silage/ water
Aflatoxins	200 g of suspected feed sample kept dry and cool
Ammonia	Whole blood, complete stomach contents and urine, should be frozen. For rumen contents, 1 or 2 drops of saturated mercuric chloride is used instead of freezing.

Poison	Materials to be sent
Ethylene glycol	10 ml of serum, whole kidney in 10% formalin and 10 ml urine
Fluoroacetate 1080	Kidney, liver, urine frozen, stomach contents and heparinised or citrated blood
Phenol	Stomach contents / rumen contents
Oxalate	Fresh plants frozen, kidney and serum
Organophosphorus compounds	Whole blood- 1ml (heparinised) sent quickly. Estimation of cholinesterase especially in nervous system, muscle or liver is helpful / rumen contents, kidney, body fat, urine, feed and hair
Rodenticides	Rumen/ stomach contents, liver, whole blood and urine
Herbicides	Rumen / stomach contents, urine and forage
Phosphorus	Liver, stomach and intestinal contents and 30 ml of oxalate blood.
Sodium chloride	Liver, serum and whole brain for histopathology
Carbon tetra chloride	Rumen/ stomach contents.
Carbon monoxide	Whole blood sent within 4 hours
Fertilizers	Stomach / rumen contents, whole blood
Chlorinated hydrocarbon	Fat 100g, liver, brain and kidney each 50 g, stomach / rumen contents 100 g and whole blood 10 ml.

Primary Poisoning vs. Secondary Poisoning

Identification of crop or stomach contents may help determine whether a case was caused by primary or secondary toxicosis. Gross comparison of ingested tissues with other carcasses found in the vicinity may help determine the sequence of toxicosis. It may be necessary to separate the contents from different organs (i.e. crop from proventriculus) and analyze portions or food items separately.

Toxicological Analyses

There is no effective broad-screen method for identifying toxins. The range of possibilities should be narrowed by the history, physical signs, use of pesticides in a given region, and previous cases in an area.

Some compounds break down relatively quickly, even those stored in a frozen state. Thus, chemical analyses should be conducted as quickly as possible. Separation of gastrointestinal contents and individual analysis on each sub-component is preferred. This measure will avoid diluting the toxin, which may occur if a composite sample is submitted.

Brain cholinesterase activities can be determined as a screening test for anti-cholinesterase toxins (carbamates and organophosphates). Brain cholinesterase suppression of 50% or greater is considered diagnostic for death by anti-cholinesterase poisoning; however, levels frequently exceed 70% under experimental conditions. Wide variation occurs in brain cholinesterase activities under field conditions. This may be confounded by the absence of "normal" reference values for many species. In addition, animals exposed to carbamate pesticides may undergo spontaneous reactivation of brain cholinesterase activities. Carbamates are reversible inhibitors, while OP's are non-reversible inhibitors. Thus, animals that succumb to carbamate poisoning may have brain cholinesterase activity within reference limits. This biochemical characteristic has been used to develop additional tests that help distinguish between carbamate and OP exposure.

Interpreting Results

Following exposure to a toxin, many processes may affect subsequent post-mortem analyses. The amount of a toxin quantitated in the laboratory is what is left following absorption into tissues, dilution in ingesta, and decomposition of the compound. Also, vomition or regurgitation will diminish the amount of a compound remaining in the carcass. The lack of poison-laced ingested material does not preclude toxin exposure, as water and percutaneous exposure also may occur. For these reasons, thorough evaluation of toxicosis cases includes ruling out other possible causes of death.

Poisoning

Precautions

Always remember that human exposure is a risk with any possible toxicosis case. Adequate protective wear and disposal of tissues and materials is critical.

Pesticides and Poisons - Pesticides and poisons are often used illegally to deal with 'pest' animals, or kill a difficult –to find predator, the latter situation often involving a 'bait' animal or a food product laced with the chemicals.

Note: Avoid using glass containers to package and ship pesticide and/or poison evidence items, as the containers may break during transit.

When investigating poisonings, consider

- Agricultural practices in the area
- History of previous poisoning incidents
- Local pest control activities
- Locality and distribution
- Epidemiological profile
- Clinical signs
- Variety of species
- Predator/prey relationships and
- Who is authorized to use such products in that area

The crime scene investigator should consider

- The source of the poison
  - Ingestion
  - Contact
- If the victim was the primary or secondary target
  - Primary consumer of poison bait

- Consumed another animal which was primary target of poison .

Documenting the poison route involves

- The sequence of ingestion
- Identification of food items
- Analysis of individual items
- Consideration of water versus cutaneous (absorbed through the skin) route
- Determining time delay and travel from the potential source of poison
  - Personal protection

Always wear protective equipment (gloves, mask, coverall suits to go over clothing, booties, etc.).

If the carcass smells "chemicals", the exposure to pesticide may be unsafe.

- Do not transport baited carcasses inside a vehicle or in the trunk.
- Do not contaminate camera, writing utensils, or other equipment.
- Double glove so that the outer contaminated glove can be removed if it develops a hole.
- Pesticide and poison collection kits are available from forensic laboratories that will conduct the analysis of the collected samples.

8 | Record Keeping

The necropsy report of a forensic case is different from a standard necropsy. A major difference between an ordinary necropsy of wildlife disease diagnostic laboratory and a forensic necropsy relates to the completeness of records that are kept about the procedures, observations and the results obtained. Most forensic cases reach court long time (months) after the necropsy was completed and several years may elapse before the time the pathologist is required to testify. Thus, if necropsy findings are documented meticulously and readily after examination, it will be easier for a pathologist to recall the details about a case should his/her testimony be required.

- The forensic investigator should follow a critical record keeping system

- The necropsy report, photographs, ancillary diagnostic reports, etc. all should be easily linked and traceable to a common source and withstand scrutiny.

- The data given on the evidence tag, name of the submitting officer and agency should be noted in the report.

    The identity of the specimen (with photographs), history, age, nutritional state, concurrent disease or injury, etc. should be documented. . These findings are relevant in cases where the cause of death is debatable, such as in toxicoses. Significant findings should be photographed with the marker throughout the entire necropsy. Use photography to preserve visual evidence. Any evidence collected, such as projectiles or tissue samples must be given their own numbers and the chain of custody continued as these samples are sent to other laboratories or experts. .

A detailed, clear necropsy report, written in non-technical language that will be easily understood by the non-medical personnel who will have to deal with it in court should be prepared. Special precautions are required for collecting specimens, such as bullets, suspected toxins, and material for DNA analysis, and for sending these specimens to other laboratories

The pathologist who appears in court long time after performing postmortem examination, may not remember the details of an individual necropsy. He depends on the written report. The pathologist should first ensure that complete history is recorded and the identity of the specimen or the animal submitted corresponds to that specified in the history. The pathologist should be fully familiar with the circumstances in which the animal was found and the nature of problem that is suspected by the enforcement officer. The pathologist should ensure that all features and results of the examination, including the history submitted with the animal, descriptions, diagrams and photographs produced during the necropsy, the results of subsequent ancillary tests, (such as histology, microbiology and toxicology), and physical specimens are clearly linked together in the record-keeping system used. This is done by assigning a unique number to each case and then using that number in all matters related to the case. If more than one animal is involved, individual numbers should be assigned to each. The pathologist must be scrupulous in recording the connection between any identification applied by the enforcement officer and the identifying number used in the laboratory. Before beginning the necropsy, the pathologist should make a final check to ensure that all numbers correspond. The final necropsy report will form the basis of any examination that occurs in court. The animal's sex, age, and body condition should always be recorded, together with the criteria used to establish these facts. Data given on the evidence tag should be noted in the report including the name of the submitting officer and agency. All descriptions should be concise using non-technical terminology when possible. All conclusions should be supported by relevant information and alternatives discussed when results are equivocal. All organ systems should be examined and all abnormalities described. Normal findings should be described as well. These extra steps provide written support that a complete and thorough necropsy was performed.

Record Keeping

Negative and positive findings should be recorded. The abnormalities and lesions should be described in absolute units (cm/grammes) rather than in relative terms like enlarged, smaller than normal. The location of injuries and abnormalities should be measured and recorded in relation to the readily identified anatomic land marks. Outline drawing are useful for recording the location of lesions such as traumatic wounds. These form the permanent record of the case. Photographs are invaluable and useful in the court for explaining the nature of lesions. If possible the pathologist take photos or be present when they are taken. It is desirable to take photograph of intact animal including its identification tags, before beginning the necropsy. Every photo should contain information that clearly identifies the subject animal, the date, the size reference scale and the identity of the pathologist. Markers or symbols should be placed on the specimens to indicate area of interest such as bullet wounds. Each photo should be accompanied by a clear description of what is illustrated, prepared by pathologist when it is turned over to the EO. Rough notes or dictation made during the necropsy should be used to prepare necropsy report immediately after necropsy is completed, since any delay may result in details being forgotten. Original notes should be retained by the pathologist until all possible court proceedings are completed. If ancillary tests are to be done, a preliminary report is prepared that includes a list of samples retained for analysis as well as the list of radiographs and photographs taken. The final report should contain the pathologist's observations and the results of ancillary tests, a list of diagnoses, a list of all specimens taken and their disposition as well as pathologist's interpretation of the meaning and significance of findings. If there are inconsistencies or if some results are equivocal this should be pointed out and alternative explanations should be discussed. After typing, final report should be read carefully and signed in each page by pathologist before it is sent. The report should be given to the authority only(confidential) who requested the necropsy. All documentation relating to each case that is not returned to the EO including the copies of the necropsy report, photos and results of ancillary tests and physical specimens must be retained by pathologist in a secure manner until the case goes to the court or until the EO advises that the case has been closed and the information is no longer required.

The pathologist should review the documents thoroughly and be familiar with the case before attending the court. One should take copies of all records pertinent to the case to the court. The pathologist must be able to differentiate between the observations made and the inferences drawn from the observations in the case. He must be careful while defining those matters where conclusions could not be drawn (

often the biological material will not yield results which the police or court requires or yield with desired degree of precision). Finally the forensic necropsy is tedious and require rigid attention to detail and may become a common part of work of a wildlife pathologists.

9 Wildlife Crime Investigation

Wildlife crime is a growing international problem that threatens the survival of many species. Combating wildlife crime is essential to preserve our natural heritage. The wildlife crimes fall in the category of serious crimes punishable by an imprisonment of at least 4 years or more.

Enforcement Agencies

- State Forest and Police Departments
- Central Bureau of Investigation (CBI) with the permission of State Governments concerned.
- Wildlife crime control Bureau (WCCB) selected wildlife offences with trans-border ramifications for investigation.
- Customs Department and the Directorate of Revenue Intelligence (DRI)
- The border guarding agencies and CISF
- GRP, RPF and Postal authorities

Hunting and illegal trade are the major wildlife offences. All other offences like preparation, possession, transportation, processing etc are ancillary offences. Based on this premise wildlife offenders can be divided into two groups - (a) the poachers or hunters who kill or capture wild animals or collect wild plants and (b) persons buying hunted and/or captured animals or its body parts or derivatives or collected plants or its parts or derivatives, for own consumption or for trade. Poaching is often associated with different levels of violence.

Linking the suspect and victim in an illegal hunting is difficult and frequently complicated by (1). The circumstances under which an animal can be legally killed, (2). The lack of species-specific definitions for wildlife parts and products and (3). The environmental conditions that act on the scene

1. The circumstances under which an animal can be legally hunted and killed are often quite complex. One must know the following information.

a. How was the animal hunted and killed?

b. The type of weapon used (Legal hunting may be restricted to the use of specific types of weapons (i.e.: archery, black-powder firearms, etc.) to kill certain animals during a defined hunting season, or within a defined hunting area.

c. When did the animal die. ? It may be illegal to kill migratory birds before sunrise or after sunset during a defined hunting season, or illegal to kill most animals the day before a defined hunting season.

d. Where was the animal killed? It may be legal to hunt in a government protected wildlife refuge during a certain hunting season, but illegal to do so in adjoining private property.

e. How many animals were killed by the hunter during particular hunting season? It may be legal to kill one animal of a specific species during a defined hunting season, but not more than one.

f. What are the genders of the animals killed? For example, it may be legal to kill a male whitetail deer during a defined hunting season but not a female.

g. It is difficult to tell which hunters killed 'over their limit' and which hunters hit nothing at all in a typical waterfowl hunting blind situation where several hunters may be shooting in the same area and the dropped ducks tend to be comingled.

h. Was the animal killed in self-defense? It may be legal for a hunter to kill an other-wise completely protected endangered species in self-defense if the animal was charging.

Wildlife Crime Investigation

2. The Lack of Species-Specific Definitions for Wildlife Parts and Products.

In wildlife investigations, the collected evidence items are often parts and products of animals from a wide range of endangered, threatened or protected species of animals from boats, planes, vehicles, residences and warehouses. The investigators rarely know the country of origin of the items and cannot depend on those who do know (usually, the suspects) to provide a dependable answer. They rarely seize whole animals as evidence. Many wildlife crime investigations will rely on evidence from professional or expert witnesses. The court may be best served by using a veterinary pathologist, with experience in a particular animal group, who is able to provide specific expert evidence. The choice of specialist or expert may need to be considered at a very early stage in an investigation. The proper collection of evidence and avoidance of potential cross contamination is an essential starting point in any forensic investigation.

When submitting items to other agencies and individuals the following should be kept in mind.

- Can the agency/individual undertake the work to a desired standard?

- Can continuity and security of evidence be maintained throughout the case? Are they aware of any potential cross contamination problems?

- Do they understand the need for unique labeling of exhibits?

- If they intend to take photographs have they been given appropriate advice on labeling, and how to handle film or digital images to ensure continuity of evidence.?

- Are they aware of evidential issues such as unused material and the need to retain all notes, documents etc? Creating a file to hold all case information is a sensible system.

- Is the person undertaking the work prepared to attend court and would they make a suitable witness? Are they satisfied they can respond appropriately to cross examination or defence evidence, Are any health and safety issues covered.

- Careful planning, systematic and meticulous examination, coupled with the use of the correct and properly calibrated equipment, deployment of appropriate field techniques and the selective application of modern technology.

Effective investigations require a combination of portable and easy-to-use laboratory equipment coupled with modern methods of data collection and information transmission. An interdisciplinary approach is essential. Biologists/naturalists and those experienced in health studies, especially epidemiology, can often usefully complement the role of the police, enforcement officials and crime scene specialists. The identification of species of animal and of other relevant evidence, e.g. ectoparasites, carrion beetles, animals and plants in the gastro-intestinal (GI) tract may need an input from people from a wide range of backgrounds – biologists, naturalists, bird-ringers (bird-banders), pest control officers, botanists, staff of museums or zoos.

In all wildlife work

Avoid the following actions at the crime scene.

- Treading or walking carelessly: Damage to footprints, type marks, vegetation

- Not wearing gloves: Addition to fingerprints

- Wearing gloves but not taking care: Obliteration of fingerprints

- Failing to wear adequate protective clothing: Contamination of crime scene with e.g. foreign DNA, Exposure to pathogens, poisons etc

- Smoking: Contamination of crime scene with ash etc

- Using domestic facilities at the site such as toilets, wash basin or telephone: Contamination, Addition or obliteration of fingerprints, Destruction (by flushing toilet or running water) of evidence.

- Moving items or turning on / off taps or lights: Evidence that animals have not been fed, watered or kept warm/cool may be lost or damaged if the facilities are touched or used.

Methods used in wildlife crime investigations

- Scene of crime visit and assessment.

- Investigative interviews.

- Collection and identification of specimens, including derivatives and specimen for laboratory testing.

Wildlife Crime Investigation

- Clinical examination of live animals.

- Post-mortem examination of dead animals. Correct storage and dispatch of specimens for laboratory testing.

- Laboratory investigations.

- Production of report(s).

The following are usually helpful in a wildlife crime scene investigation

- Accurate information about the circumstances of the alleged offence (a 'history'), preferably in the form of a written record.

- Discussion with those involved in reporting the incident, initiating the investigation or in bringing charges. This, too, should be covered in a written statement wherever possible.

- Correct identification of the species of animal(s) involved.

- A site visit to the location where the offence is alleged to have occurred (the 'crime scene') or the premises where live animals need to be viewed or dead animals/their derivatives can be examined.

- Immediate recording of environmental parameters, including weather data.

- Proper collection, holding, transportation, storage and submission of evidence and trace evidence and the relevant accompanying statements and records. A problem when visiting crime scenes is that 'contamination' — the introduction of extraneous, irrelevant, material — can easily occur.

- Careful documentation and record keeping, including appropriate photographic records, leading to production of a report. Ultimately, if the matter proceeds, it will be necessary to prepare evidence for, and possibly appear in, a court of law.

Assessing whether a criminal offence has taken place may not always be straight forward. Other possibilities such as natural deaths, predation and legal hunting should be considered. Care should be taken not to cause cross contamination problems in relation to samples taken from a crime scene and dealing with a potential suspect. At the outset consider all evidence types, utilization of experts and the order of examination. Protect the Wildlife crime scenes. In some cases there may be value in cordoning off an area or creating an exclusion zone to minimize the chances of evidence being damaged or destroyed. Minimize unnecessary visits. Have an agreed access route (also referred to as a common approach path) for those needing to attend the scene. The scene may need to be visited by the investigating officer, perhaps in company with a reporting witness, a CSI officer, a badger expert and possibly other personnel. CSI officers should, have precedence over other parties. There may be value in the CSI officer being directly accompanied by individuals with specialist knowledge so that specific items can be drawn to their attention to record, collect and sample.

The holding of evidence can be particularly problematic in wildlife cases, especially under field conditions. The holding of live animals that are evidence or confiscated as 'seizures' can present major difficulties, as these will need to be properly housed and provided with adequate care.

Coordinating work in the field can be difficult, especially if Scene-of-crime officer. (SOCO)s or their equivalent are unavailable to assist.

Like wise establishing and maintaining a crime scene can present problems if those first at the site have limited experience of such matters. Specialists in crime scene management together with police officers, staff of enforcement bodies and others, including those with experience of the species in question, usually have a major role to play in investigations. **Two** major groups of people viz. (1) amateurs (field naturalists, bird-ringers-bird banders, game-keepers and trackers) who often have detailed knowledge of the location and who can assist in such tasks as the recognition and identification of animal footprints, hair, feathers, droppings and (bird) pellets; and (2) medical and veterinary personnel experienced with epidemiological field studies can often assist the investigation. In addition ecologists, entomologists, botanists, veterinarians, pathologists, molecular biologists and toxicologists should be incorporated in forensic wildlife work.

Identification of animals or their products in forensic cases

There is always a need for accurate identification of animal products or derivatives. Sometimes the product is immediately recognizable as coming from a particular taxon of animals, e.g. snakeskin, deer antlers, bird eggs. However, it is not easy to recognize when blood, bile or other body products are seized. If there is a suspicion

Wildlife Crime Investigation

of illegal use or disposal of chemicals at the crime scene, the naturalist may notice the dying caterpillars or unusual numbers of fish on the surface of a lake and provides the important initial clues that can lead to a prosecution. The naturalist has broad knowledge of taxonomy, behaviour and animal biology. Correct identification of taxa is always a key part of a forensic investigation and appreciating how animals live and behave can play an important part in elucidating the sequence of events in a wildlife case.

In a typical case of wildlife crime, the investigating officer should plan investigation in the following manner

1. Investigation at the scene of crime/scene of occurrence/Post Mortem of Carcass

2. Interrogation of the accused(s)/suspect(s)

3. Examination of the witness(es)

4. Collection of documentary evidence(s)/samples for expert opinion.

5. Collection of scientific/forensic evidence/digital evidence(s)

6. Collection and analysis of the evidence(s) and

7. Filing complaint u/s 55 of the Wildlife (Protection) Act, 1972.

Wildlife crime scene investigation (CSI) Team

It is desirable to have an experienced investigative team at the crime scene, each having specifically assigned tasks and appropriate expertise. Whenever possible, it is advisable to have a planning meeting which gives an opportunity to bring everyone together and to minimize the risk of inadequate communication. One person has to be in charge and brief the investigative team but everybody must have an opportunity to speak and to comment. A systematic plan will ensure that all those involved are aware of the steps that will be taken, that the scene of the alleged crime is fully examined and that potential evidence is not lost or missed during the process. It is essential to prepare well for site visits and fieldwork and to ensure that those involved are familiar with the necessary rules, standard operating procedures (SOPs) and codes of practice. This is particularly important when backup or direction from the police or other experienced people is not available.

Role of each member of Investigative Team.

Leader (1)	1.Contact the first officers at the scene 2.Review the area for safety concerns and inform the rest of the team 3.Conduct the initial scene walk-through (with the evidence collector) and 4.Communicate findings to the rest of the team 5.Determine the setting of the scene perimeter 6.Take note of weather conditions and time of day (it may be better to secure the scene and wait for daylight or better weather) 7.Determine the number of persons/abilities needed (such as a diver) and additional equipment needs 8.Assign CSI tasks to team members 9.Be the point of contact for media releases 10.Update superiors as to the status of the scene investigation 11.Monitor and record who comes in and out of the scene 12.Oversee the completion of all CSI tasks 13.Determine when all CSI tasks are completed

Evidence collector (1)	1.Accompany the team leader during the walk-through 2.Place flags next to located evidence items during walk-through and scene search 3.Place numbered evidence location tags next to evidence items 4. Prepare the located evidence list 5.Collect all evidence items 6.Mark, package and tag all collected items

Wildlife Crime Investigation

Scene photographer (1)	1.Take initial scene photos from outside the scene perimeter 2.Take 'overall' scene photos on the initial walk-through (showing the scene 'as found') 3.Take 'overall', 'mid-range' and 'close-up' photos within the scene perimeter of all evidence items 4.Take photos of any animal carcasses, preferably at right angles to the body alignment 5.Photograph all located impression marks (footprints and tire marks) 6.Take 'close-up' photos of all located evidence items
Scene sketcher (1)	1.Prepare the rough scene sketch at the scene 2.Indicate North, South, East, West directions on sketch if known 3.Show rough location of victim animal(s) on sketch 4.Show rough location of all fixed reference points used for three-point measurements and evidence items on the sketch 5.'Map' the exact location of evidence items as appropriate with 'distance-to' Measurements. 6.Prepare a finished sketch for presentation in court
Remaining team members to assist in	1.Scene security 2.Scene search 3.Evidence mapping-Place flags next to evidence items. 4.Assist the scene sketcher in mapping the exact location of evidence items using a tape measure and 'fixed' or relatively permanent objects at the scene as measuring points 5.Collect flags and scene perimeter tape after CSI tasks are completed

It is usually performed by one or two investigators due to scarcity of personnel.

Techniques

The basic steps in a wildlife crime scene investigation are

1. Arrival at the scene

2. Noting and addressing any safety concerns

3. Setting initial scene perimeter

4. Scene security

5. Initial scene photography

6. Initial walk-through

7. Preservation of fragile and easily lost or destroyed or altered evidence

8. Re-assessment of equipment or additional personnel needs

9. Briefing of other scene processors

10. Scene search (this continues throughout the scene processing)

11. Visualizing the location of evidence items at the scene

12. Evidence location tagging

13. Evidence location list

14. Crime scene photography

15. General CSI photography guidelines

16. Mapping the location of evidence items

17. Crime scene sketch

18. Marking evidence items

19. Packaging and sealing evidence items

20. Methods of sealing evidence packages

21. Tagging evidence items and packages

22. Evidence receipts

Wildlife Crime Investigation

23. Chain-of-custody record

24. Evidence submission form

25. CSI report

26. Specific evidence collecting/ packaging issues

- Pesticides and poisons
- Blood and tissue
- Firearms
- Bullet and cartridge casings
- Tool marks
- Friction ridge evidence Footwear and tire tracks
- Trace evidence
- Animal traps

Wildlife Crime Investigation Kit

1. Surgical Gloves — - 2 Pairs

2. Plastic Pouch (Cellophane) - 4 Nos.

3. Screw capped vial (50 ml) - 4 Nos.

4. Screw capped vial (5 ml) - 4 Nos.

5. Injection Syringe (5ml) - 1 No.

6. Forceps - 2 Nos.

7. Brush - 1 No.

8. Glass Slides - 20 Nos.

9. Silica Gel - 20 gm

10. Slide case - 1 No.

11. Filter Paper - 1 No.

12. Measuring tape - 1 No.

13. Scissors - 1 No.

14. Magnifying Glass (Small)   - 1 No.

15. Cell Flashlight (torch)   - 1 No.

16. Marker   - 1 No.

17. Eye Protector   (orange/blue)

18. Notepad

19. Pen/Pencil

20. Ruler

21. Pocket Knife

22. Foul Weather Gear (Raincoat/Umbrella)

23. Waterless Hand wash

24. SLR/Digital camera

25. Preservation Kit for Blood Samples

26. Carry Bag

27. Light Source with Filters

28. Crime Scene (Protection) Strip

Pug Mark Collection Kit

29. Tracer   25 cm X 20 cm glass/ plastic with 2cm broad Wooden frame

30. Sketch Pen   -1

31. 2m long measuring tape

32. Plaster of Paris

33. Water Bottle   1 litre

34. Flexible Aluminium Tape

35. Census Bag   To keep the items (19-20)

Wildlife Crime Investigation

Working in the field.

The amount of forensic investigation that can be performed at the crime scene depends upon

1. The location of the crime scene and its accessibility,

2. The type of investigations that are needed and

3. The equipment and personnel (both at the crime scene and back in the laboratory) that are available.

In high profile cases, safety, security and press (media) interest have to be taken into account. Some forensic investigations can only be carried out adequately in the field (examination of animal footprints in sand or in mud, the temperature or smell of the faeces). Others can be performed in the laboratory (faeces can be packaged and transported to a laboratory). Each situation has to be judged on its merits.

Equipment

- Compile a checklist of items to be taken.
- Always include spectacles, medication (a comprehensive first-aid kit) and other domestic items that can make life in the field more comfortable. One must be cautious about substances (e.g. mosquito repellent) that might contaminate evidence.
- All items must be scrupulously clean. This should be documented and may become evidential, e.g. where contamination by investigators is alleged. On leaving, after checking that evidence has not been removed, everything must be thoroughly cleaned and decontaminated.

a) Specialized equipment

i. It is advantageous to have an **ornithologists and other naturalists** in wildlife crime scene work, since they have relevant **equipment** (such as that used for bird watching like waterproof and submersible cameras and cases, night-vision kits, monitoring devices, GPS equipment and hand-held radios.) that can readily be adapted to forensic studies.

ii. If the investigations are likely to involve the capture and handling of animals, specialized items may be needed (gloves, nets, bags, capture and restraining devices, binoculars, reference texts, avian ring (leg-band) pliers and remover, a magnifying glass and scales or balance).

b) Portable equipment (microscopes and endoscopes) that can function mechanically, on solar energy or by using batteries into difficult terrain (working far from roads or ready means of communication, with little or no access to clean water and without mains electricity). The construction of well-defined areas in forests, deserts, or elsewhere, to conduct necropsies efficiently and safely and carrying a cell or satellite phones or radios for communication are essential.

c) Think of a plan to deal with natural or anthropogenic dangers.

d) Portable field kits to collect samples for Modern scientific methods (DNA-based).

e) Microscopes and other equipment that usually runs on electricity can be used in the field.

f) Battery operated equipment viz. miniaturized otoscope (auriscope), ophthalmoscope, rigid endoscope, colorimeter, electrocautery, blood-pressure monitor, respiratory monitor and pulse oximeter, minicentrifuge, miniphotometer, and refractometer. Lightweight flexible solar panels. They can be folded or rolled and inserted into protective tubes for Transportation to the field.

Some techniques used in the wildlife crime investigation

1. **Gross and Microscopic studies**, isoelectric focusing and DNA-based techniques are used for species determination using bones, skins, hair, feathers, scales and other organs and tissues/ meat. Detailed Scanning Electron Microscopic studies can reveal important taxonomic features. Electrophoresis including gel diffusion studies of tissues and meat are used for species identification.

2. **Examination of whole carcasses**, organs, tissues from dead animals using radiology, scanning techniques, cytology bacteriology etc, and other laboratory investigations for Pathological studies

Wildlife Crime Investigation

3. **Examination of clinical materials from live animals** and conducting supporting tests, as above, and others-Clinical work. Live animals should, be monitored using non-invasive or minimally invasive methods.

4. A very important component of most investigations **is using DNA technology for Species, sex and parentage determination**. Karyotyping is employed using various tissues for determination of species and sex (requires normal chromosomal patterns). Also detects chromosomal abnormalities. The various DNA techniques using blood, saliva, other body fluids or tissues are used for identifying the species and individuals, parentage and sexing. These samples must be handled and stored well to prevent damage and contamination. Polymerase chain reaction (PCR)permits forensic identification of tiny, less than optimal samples.

5. **Analysis of stable isotopes, trace elements, chemical residues** and soil or sediments to get background information on the history and health of an animal, including its diet and its movements. The services of forensic geoscientists are also used..

6. **Standard 'crime investigation' involves studies on ballistics, debris from explosions, fires, fingerprints, blood spatter, tyre etc.,** plus tests on items such as non-ballistic nets and traps. These investigations areusually performed by police and specialists. Non-marks, ballistic weaponsare used to kill (deer) or maim wildlife in many parts of the world. In poorer parts of the world arrows, spheres and catapults may be standard means of taking animals.

Identification and morphological studies - In all wildlife crime investigations there is a need for accurate identification of animal products or derivatives. Sometimes the species or its parts (derivatives) are readily recognizable as coming from a particular taxon, e.g. a snakeskin, a dead parrot, antlers of deer based on morphological studies — for example, examination of bones or hair coupled with such DNA technology as PCR and sequencing, to identify certain especially cryptic taxa. When tiny portions of muscle, blood, bile or other body products are seized, DNA analysis is likely to be essential. Identification of whole animals or their parts is facilitated if a range of photographs is taken at the outset. This is particularly true of species such as cetaceans, amphibians and fish, which will rapidly change in morphology as they decompose, become bloated or are scavenged. Photography should be accompanied by morphometrics (weighing and measuring). It should be followed by the collection of gross specimens as evidence. Molecular diagnostics permit the detection and differentiation of substances from different species and provide information on genetic diversity that can assist in the conservation of endangered animal species. PCR is particularly valuable in forensic investigation as it enables tiny samples to be identified. Morphological methods of identification and additional molecular-based approaches must be an adjunct to it.

The ideal substrates for forensic DNA analysis are fresh samples of tissue (blood, muscle, etc.) collected using sterile techniques from living or recently dead animals and preserved at $-20°C$ in a dedicated freezer.

Pathological studies - Some investigations in the field seek to find out the circumstances of death. Whenever possible Post-mortem examinations should, be performed by an experienced comparative pathologist. The detection of lesions, wounds, bruises, abscesses and scars can be pivotal in determining how and when a wild animal died, or its health status prior to death. Samples need to be taken in addition to photographs as supporting evidence in court. (Digital imaging enables the sending of pictures to experts in other parts of the world). Post-mortem change (autolysis) can hamper investigation of carcasses but, if properly investigated and analyzed, can provide useful information on when the animal died and the environmental features at the time. Gross and histological examination are often coupled with the investigation of invertebrates that are present within, on, or in the vicinity of the animal's body ('forensic entomology).' Wild animals killed or injured by humans need to be differentiated from those attacked by predators, e.g. a free-living raptor or mongoose or a domesticated dog or cat. Some mammalian predators can be responsible for mass killings of ground-roosting or nesting birds. In such situations, detective work, including detection of tracks will be needed to determine the sequence of events. Tooth, beak or claw/talon marks assist in determining which predator killed the animal, as may the scene of the killing — for instance, whether portions of the prey are scattered or cached. Post-mortem scavenging needs to be distinguished from ante-mortem predation. In investigations where carcasses need to be collected for examination, scavenging and disturbance by wild, feral and domesticated animals can hamper collection of bodies and make counting difficult. The use of dogs may increase the rate of recovery of carcasses. But dogs will damage and destroy other evidence. Traps, nets and snares are often used to catch wild animals which is not necessarily, per se, illegal. Different species are trapped because they are pests, or for

Wildlife Crime Investigation

food or as part of scientific study.

Traps, nets and snares can cause characteristic traumatic injuries, as well as stress and pain. Trauma is usually characterized by 'patterned injuries.' Traps produce crushing lesions; nets and snares cause strangulation or impairment to blood flow. Poisoning of wildlife (including invertebrates) is widespread and its investigation necessitates the services of scientists of different disciplines. Acute poisoning cases are usually most likely to be detected in the field because animals are found dead or dying.

Forensic entomology - A wide range of invertebrates can contribute to forensic studies. Invertebrates may be transported on an animal's body or be found in its bedding or container. Collection of invertebrates for study and as evidence requires strict adherence to well-established protocols. These include standard operating procedures to ensure that, specimens are kept fresh (alive) for taxonomic studies and for possible rearing to adult stage as well as being fixed in alcohol. Maintaining a chain of custody is as important in forensic entomological studies as it is in any other field of forensic work.

Submission of evidence - This is critical to the success of a case. Samples must be documented, properly packed, labeled and transported, with attention throughout to both ensuring the quality and integrity of the material and maintaining the chain of custody.

Reference material - Prepare and catalogue materials collected from animals (e.g. feathers, hair or bones) properly. This has a great value in the investigation of wildlife crime. In addition to their use as sources of reference in court cases, they are important aids to teaching and research.

Record keeping at the crime scene

i. Notebooks for written information are essential.

ii. These can be supplemented with a micro-cassette tape recorder, which can be used for dictated notes and to record sounds of interest. Inserting the tape-recorder into a sock or enveloping it in a cloth or a thin polyurethane foam sleeve will minimize wind noise.

iii. Video recording is another useful way of assembling information at a crime scene. It should, be supplemented with photography, tape recording and the writing of contemporaneous notes. A Laptop can be invaluable.

1. Photography - Taking a high standard Photographs and examining dead animals and other sensitive material at the crime scene is of great importance. Digital cameras are invaluable.

A crime scene visit/ site visit to the location is the prelude to (1) collection, transportation, storage and submission of evidence; (2) documentation and record keeping, including the taking of photographs (3) production of a report; and (4) (if the matter proceeds) preparation for and the presentation of evidence in court.

2. Access the crime scene as soon as possible - The 'scene' may encompass several countries and include locations ranging from tropical forest, isolated coastline, desert or tundra (where animals have been hunted), to establishments such as the premises of an animal dealer, a traditional market, a zoological collection, or a private home. It may even comprise websites on which live or dead species or their derivatives are offered for sale. It may also involve different physical locations where wildlife crime needs to be investigated and how best to perform well under adverse conditions. The crime scene' may range from the carcass (or parts) of an animal to terrain that contains a topography as varied as forest or desert and may include diverse natural and man-made structures. Often, the location is isolated, with few facilities for proper investigation and collection of evidence. If delayed, the evidence might get destroyed or spoiled due to rain, wind or other climatic factors, or by the activities of humans, domesticated livestock (e.g. grazing cattle) or wild animals (including scavengers).

3. Keep ready the equipment (protective clothing to specimen containers).

4. Establishing a wildlife crime scene - The person who is coordinating activities must ascertain the size and location of the crime scene, (using GPS) and establish a proper command structure and responsibility by introducing him/herself to others present.

i. Upon arrival at the scene, the Investigator should assess the following.

Wildlife Crime Investigation

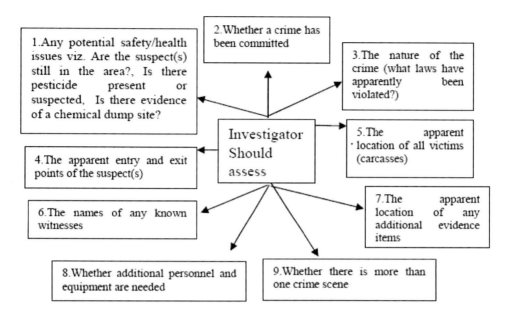

Note: If other wildlife officers arrived at the scene first, and have already approached the victim(s), the wildlife scene investigator should take the same path that the other officer(s) took in approaching the victim(s) to prevent the accidental destruction of related evidence items such as footprints or tyre tracks. Take elimination footwear, and record fingerprints and tire track information of responding officers if needed.

ii. Investigator should note and address any safety concerns of anyone visiting the location.. Apart from physical hazards imposed by the terrain, weather and lighting conditions, with wildlife incidents there may be a range of additional risks associated with toxic pesticides, handling live animals, zoonoses etc. Some scenes may require the use of specialist equipment and personnel.

Primary focus of crime scene investigator is to preserve and collect evidence.

The investigative team should be constantly alert for the presence of scene hazards like, (a).Predators and poisonous animals/insects at the scene, (b). Poisons used to kill the victim,(c).Discarded and/or loaded firearms. In case of loaded firearms these firearms should be unloaded at the scene, using caution to preserve possible latent fingerprint evidence on the item and the ammunition and removable magazines,(d).Set animal traps, (e).Disease vectors associated with the victim,(f).Scene topography (cliffs, holes, etc.), and (g).Weather extremes .

iii. Based on the initial information obtained upon arrival at the scene **set initial scene perimeter**

- Find out the size and location of the crime scene, (using GPS).

- Demarcate the crime scene by using barrier tape or its equivalent supplemented with markers, flags, canes and other standard crime-scene security items. At first, incorporate a larger area, and then adjust the area if necessary, enclosing all known evidence within the scene boundary keeping out onlookers, members of the media, and other law enforcement personnel who are not needed at the scene.

- Move the bystanders away from buildings that may collapse or animals that may bite and ensure the safety of the scene.

- Protect the evidence from accidental destruction. Focus the investigation on the most important scene areas. Maintain a record of entry of those persons for whom it is necessary to enter the crime scene, Do not admit unnecessarily.

- Plan collection and holding of evidence. Arrange photographic and other records.

- Maintain a proper chain of custody.

- The scene perimeter should be a visual barrier unlike a physical barrier. The most common materials used to set are Crime scene tape, Rope, Barricades, Vehicles and Natural geography.

- **Note:** No need to set a scene perimeter if the investigator is only responding to a scene in a remote location to document and collect evidence related to a found carcass (as opposed to trying to reconstruct the events of the crime, and searching for additional evidence).

- When searching vehicles and/or buildings during a raid or warrant search, the exterior of the vehicle and the exterior walls of the building serve as effective scene perimeters.

Wildlife Crime Investigation

4. Scene Security

a. Precautions at the crime scene

A crime scene has a legal significance and should be disturbed as little as possible.

Some useful rules when entering a crime scene are as follows.

i. Prepare a checklist of what you need beforehand in order to avoid entering and leaving the scene unnecessarily.

ii. Inform other people if you are initially visiting the crime scene alone. In Wildlife cases, visits to crime scenes, can take a long time and may prove hazardous. It is important that someone knows where you have gone and for what reasons.

iii. Follow the rules and instructions of Scenes of crime officer (SOCO), Crime scene investigators (CSI), police officer or equivalent if present. Minimize all disturbance to the site. Do not touch anything unnecessarily, Do not move when movement is not required, and do not put anything on the ground but carry everything you need.

iv. Remember the basic rule that observation is the all important prelude to action. Supplement this with photography, video and tape recording.

Once the investigator has arrived, he alone should be inside the scene perimeter. The primary task of any additional officers at the scene is to protect the scene from outside the scene perimeter. All others should be kept out. Date/time log of everyone that enters/exits the scene should be maintained.

Bona fide persons may include -

- Other pathologists.
- Appropriate colleagues, e.g. clinicians who were involved in the case.
- Support staff, e.g. technicians.
- Police officers.
- Officials from relevant animal welfare or conservation bodies.

- Where appropriate (usually wildlife cases), people from the local community with an interest in, or responsibility for, traditional use or custody of the species.

- Students.

b. Legal considerations at the crime scene.

i. Strictly adhere to international, national and local laws.

ii. Supplement this with protocols, standard operating procedures (SOPs) and codes of practice, (not necessarily legally binding). This will help to ensure a professional and ethical approach to the work.

iii. Consult local officials and community leaders about traditional practices and culture. Some legislation may limit what can and cannot be done. For instance, in certain countries, if investigations involve live animals, a registered veterinary surgeon (veterinarian) needs to be present to carry out specific procedures.

iv. Properly formulate Risk Assessment to minimize the dangers at the scene. Crime scene work in the field expose personnel to injuries (e.g. from rough or dangerous terrain or, in war torn areas, unexploded ordnance) and to assault by poachers, guerrillas, hostile witnesses or others. Issue personal protective clothing to those who handle or examine material at the crime scene.

5. Before making the initial scene walk-through, take a series of **wide angled Panoramic photographs** from outside the scene perimeter without people present to record and show how the scene appeared before the investigation began.

- Before beginning photography check that the camera's date and time settings are correct.

- Never erase a photograph on the memory disc once scene photography has begun, even if the photo is blurred or poorly exposed. This practice will prevent accusations that the investigator deliberately hid or erased evidence favorable to the defendant.

- Never mix investigative work photos with other non-investigative photos on the same camera card.

Wildlife Crime Investigation

6. At the onset of the scene investigation and before starting to collect and document evidence, make an **initial 'walk-through'** of the scene

- To locate as much apparent evidence as possible and to quickly preserve fragile evidence that might otherwise be destroyed.

- During the walk-through, locate items of evidence as they are discovered using brightly-colored flags (Fluorescent green, international orange, or pink show up well against a variety of background) and/or tape.

- Avoid using Red and yellow tape, flags and placards since they blend in with many types of foliage, especially during the Autumn season.

- Photograph these flag placements at mid-range (showing the evidence item, the flag, and a small amount of surrounding area) after the flags are put into place.

- Communicate all findings during the initial walk-through to the entire investigative team.

7. During the initial walk through **preserve the Fragile and 'Easily Lost or Destroyed or Altered' Evidence.**

- The fragile evidence items are, shoe prints or tyre tracks in loose soil, snow, sand or under water, loose papers, loose trace evidence (hair, fibres, etc.), wet blood, exposed surfaces with possible latent fingerprints, shoe tread exemplars from suspects, tyre track exemplars from suspect vehicles, finger and palm print exemplars from suspects, and blood samples or trace evidence from shoes and clothing of suspects.

8. Make a methodical search of a crime scene **for physical evidence items which is crucial.**

- The purpose is to make certain that all relevant evidence items are discovered during that search. Secure the scene completely in order to come back at a later day and time to find additional items of evidence in a manner that will be acceptable to a court of law.

- Use a combination of systematic and thorough search methods depending on number of people available, geography etc. It will be useful to re-search the same initial area from the opposite or different direction as lighting changes may make evidence items such as cartridge cases visible later in the day as the sun changes direction.

- During the CSI process, all personnel should remain alert to the possible presence of additional safety issues that may impact the investigation and/or the individual team members.

- An effective search of a crime scene and surrounding area will often provide important evidence. First consider the type of offence under investigation and activities of those involved. Crimes may involve people visiting an area to commit offences, such as poachers, people digging at badger sets or egg collectors or may involve local people who regularly use the land, such as a wildlife poisoning case or illegal snaring and trapping.

- Where offenders are suspected to be local land users then an initial low key visit may reduce the chance of alerting potential suspects.

- In offences such as wildlife poisoning cases, larger scale searches of land and the use of specially trained police search teams may be appropriate.

- The assistance of specialist staff who are familiar with the types of offences under investigation, and who can assist with the search effort and provide an assessment of whether items found should be seized or sampled etc. should be sought where possible.

9. Visualize newly discovered items of evidence like a pile of expended cartridge casings and/or objects of interest like fresh bullet holes in a nearby traffic sign (the sign will not be collected, just documented) as to location with bright flags or tape. Identify the additionally identified and collected items like the expended casings with numbered signs. Once this is done, take a few 'over-all', 'mid-range' and 'close-up' photos to show the distribution of located evidence items and objects of interest at a crime scene.

10. Provide unique identifying numbers to evidence items and objects of interest at a crime scene using Evidence Location Tags and also to keep track of work that needs to be done at the crime scene.

Wildlife Crime Investigation 297

- Write the case (investigation) number, the date, the investigator's initials and a sequential Evidence Location Tagging (ELT) number on torn pieces of a grocery bag.

- Place the ELTs inside or attach to the evidence packages with the items to aid in later identification of those items in court.

11. Make a list of all objects of interest and evidence items documented (by a photograph and a sketch) and/or collected at a specific crime scene.

- The Located Evidence List includes Investigation case number, Date of evidence collection, Location where the evidence collection took place, Name(s) of investigators involved, Evidence location tag numbers, Individual evidence item descriptions, Individual 'objects of interest' descriptions, Individual evidence item locations, Individual evidence items found by, Individual items collected by, and Collected Item numbers or seizure tag numbers.

- It is better if one investigator collects, marks, packages and tags all of the located evidence items at a wildlife crime scene.

- Incorporate the located evidence list into a more comprehensive crime scene investigation (or evidence collection notes) report.

12. Crime Scene Photography

- Before taking photos at a wildlife crime scene check the date and time settings on the digital camera.

- Adjust the image size to match the size and pixel density of desired photographs and the storage capability of memory card. Format all memory cards that will be used for the scene investigation.

- Take Over-all photos of the entire scene (some shot from outside the scene perimeter looking inward), Mid-range photos showing one or more evidence items or objects of interest, and including some surrounding area for reference and Close-up photos of each evidence item or object of interest (including an ELT or other placard) and also including an L shaped ruler for footprints, tyre tracks and an individual number for each area of friction ridge detail (finger/palm prints).

- Document photos showing the evidence items, Evidence photos of finger/palm/footwear/tyre-track evidence that is not lifted or cast and Aerial photos using aircraft or internet services such as Google Earth.

- Take Photographs of 'impression mark' evidence (e.g.: shoe prints or tyre marks) at a 90° angle to the imprint surface (straight down) using oblique (angled) lighting and an L-shaped ruler located at the same depth as the impression detail. This will enable a forensic examiner to 'blow up' the shoeprint image from the scene to the correct size before comparing it to the shoe(s) or tyres(s) in question.

- Take impression mark photographs with the use of a tripod and detachable flash when they are available.

- After photo-documenting the tyre tracks or footprints, make casts with dental stone or similar material to preserve the impressions for court presentation.

- Re-photograph the scene during the day as lighting conditions change.

13. Guidelines for General CSI Photography -

- Take the first photograph of a paper clearly listing (i).The investigation case number, (ii).The date,(iii). The scene location, (iv).The photographer's name and badge number, (v). The investigating agency and (vi).The start and ending times

- Take at least one 'over-all', 'mid-range' and 'close-up' digital photo per evidence item at crime scenes.

- Avoid standing in one place and take the same shot repeatedly. Whenever possible, take crime scene photographs at right angles to the primary object of interest. This will result in the least amount of photographic distortion of the images. This can be documented by photographing the marked paper twice, at the start and at the end of the scene investigation.

- Though it is not necessary to include some kind of identifying tag or number for the initial over-all and initial walk-through photos it is always a must in the mid-range and close-up crime scene photographs.

- Do not erase digital photos taken at a scene from a memory card, even if you know they were out of focus or poorly exposed. Be careful about who or what to include in crime scene photographs.

Wildlife Crime Investigation

- Do not include, people, if it can be avoided, officers' vehicles / plate numbers, faces of investigators, items, activities and people not related to the investigation and suspects in a CSI photo .

- Do not mix crime scene and unrelated photos on the same memory card.

- Make sure that the audio setting is off and warn others at the scene if videotaping.

- It may be a good idea to use more than one camera and use more than one memory card (especially on highly important investigations). If one camera fails or a card becomes corrupted one may not lose all photos.

14. Mapping the Location of Evidence Items is done to show the precise locations of evidence items on a crime scene sketch. This allows an accurate 'reconstruction' of the crime scene at a later date. The two main variations of evidence mapping are (a) Indoor mapping: The precise location measurements for evidence items are measured at right angles from two walls, and (b) Outdoor mapping: Rough location measurements for evidence items are made from at least three fixed points, resulting in a 'triangulated' mapping of the evidence item.

Note: All measurements must be made at right (90 degree) angles to the wall.

15. Make the sketch of the Crime Scene Sketch to supplement photographs.

- A 'finished' sketch can provide more information than an 'over-all' photo about the scene.

- A crime scene sketch should include, the investigative case number, the date of making the sketch, the scene location, a brief scene description, the sketcher's name and badge number, the name of the investigating agency, direction of North, South, East, and West if known, and the words "Not To Scale", the location of all evidence items and the location of any permanent reference points used in 'mapping' evidence items.

16. Whenever possible Mark all collected Evidence Items (or at least the evidence packaging) permanently for later identification with (1).The investigative case number, (2). The unique item number, (3).The investigator's name (or initials) and badge number, (4).The date of collection.

17. Pack all the evidence items into new and clean paper envelopes, paper bags, cardboard boxes, plastic jars or vials and metal paint cans and then **seal the package** in such a manner as to prevent undetected access into the package by unauthorized persons. The purpose of a sealed evidence package is to ensure that the contained evidence items have not been switched or altered in any manner during storage and prior to presentation in a court of law.

18. General Packaging Guideline for Evidence items

- The evidence package should be strong to completely contain the evidence item without tearing or breaking. If necessary, place the first package into a protective and sealable cardboard box. Place the Evidence Location Tags (ELTs) inside the evidence package with the evidence item(s), making sure it will not damage possible latent prints. This will help the investigator identify the evidence item later in a court of law.

- Do not seal bloody evidence (clothing, weapons, tissue, etc.) into plastic bags.

- Always use paper bags.

- Transport wet/bloody items (for a short period of time, and under moderate temperature conditions) in plastic bags to a drying/holding facility.

- Freeze bloody and/or decomposing carcasses and then seal in plastic bags or ice chests.

19. Tagging Evidence Items and Packages

- Tag all sealed evidence packages and all loose (non-packaged) evidence items for

  - identification with an Evidence Seizure Tag which is attached to the exterior of the evidence package or loose evidence item.

- Provide the following information.

  - A unique seizure tag number, The agency investigation (case) number, The date and time of seizure, The seizure location, A description of the evidence item(s) and The name and badge number of the investigator making the seizure. Some Agency Seizure Tags also include a removable 'Seized Property Receipt' (which can be filled out, torn off and given to the owner of the property at the time of seizure) and a chain of-custody record (usually on the back of the tag).

20. Evidence Receipts - As a general rule, if an investigator takes custody of an item of evidence, that investigator must either 1) remain in possession of that evidence (ideally in a sealed evidence package, and placed in a locked storage area), or 2) possess a receipt for the transfer of that evidence to another person or storage facility. An evidence receipt can be A copy of a signed Chain-of-Custody Record, The torn-off receipt (signed) from an Evidence Seizure Tag, The signed back of an Evidence Seizure Tag. A separate evidence receipt form and A handwritten and signed receipt.

21. Chain-of-Custody Record

- A Chain-of-Custody Record documents the possession (custody) of evidence items from time of seizure to the time of final release or disposal.

- The original Chain-of-Custody Form (bearing original signatures) should accompany the sealed and tagged evidence packages and/or loose tagged evidence items.

- Persons releasing custody of evidence packages and/or loose evidence items to another individual or evidence storage facility should keep a copy of the completely signed Chain-of Custody record as a receipt for the released evidence.

22. Evidence Submission Form - When submitting items of evidence to a forensics laboratory for analysis, the use of an Evidence Submittal Form helps to clarify what evidence items were submitted and what examinations were requested.

23. The CSI Report - It is a compilation of all activities performed, and all evidence items located, documented and collected by the CSI team or other persons.

Note - CSI notes and reports can be made in any format (including blank sheets of paper); but pre-printed forms help ensure that all necessary information is recorded at the scene (an important consideration when investigators are tired or working under stressful conditions).

Important note - CSI reports and all other crime scene forms must be filled out in ink, never in pencil. If a mistake is made, draw a line through the error and initial the correction.

Collection of scientific/forensic evidence : Institutions from where assistance can be got.

S. No	Items	Purpose	Institute
1	Seized plants or plant derivatives	To establish the genus and species of the plant	Botanical Survey of India (BSI), Kolkata.
2	Seized living animals and Seized shells, coral etc	To establish the genus and species of the animal	Zoological Survey of India (ZSI), Kolkata
3	Seized meat	To confirm the genus and species of the animal,	Wildlife Institute of India (WII), Dehradun

24. In case of unnatural death of animals - Post mortem examination is to be done by the Government Veterinary doctor or a team of Veterinary doctors, in presence of independent experts and witnesses to establish the cause of death, weapon used and probable time of death.

For carrying out Post Mortem the guidelines issued by National Tiger Conservation Authority (NTCA) and other authorities concerned should be followed .Viscera should be collected and sent to the **State Forensic Science Laboratory** for toxicological examination in case of suspected poisoning cases for confirmation and identification of the poison used. Viscera should be sent to the **Wildlife Institute of India (WII), Dehradun**, for pathological examination in case of deaths reported due to disease.

25. Seizure of skins - Examine the skin thoroughly for the presence of chemicals used as preservatives. Collect samples with the help of forensic experts, wherever possible, and send the same to the **Forensic Science Laboratory** (FSL) to find out the **type of preservatives** used. Also send the sample to the nearest **Forensic Science Laboratory for Ballistic examination** to find out the type of fire arms used, range of firing etc. Ballistic examination will also establish presence of Gun Powder Residues on the skin. If the animal is killed by firing, the skin in case of tiger skin take a digital photograph at right angle and send to **Wildlife Institute of India (WII)** to match the stripe pattern of the dead animal with the camera trap repository available in WII, to establish the habitat from where tiger may have been poached.

Institute	Opinion offered
Wildlife Institute of India, Post Box No. 18, Chandrabani, Dehradun - 248 001,	For Identification of all animal species. Identification of blood, tissue, hair, bones, nails, claws, teeth other body parts and derivatives.
Zoological Survey of India , Prani Vigyan Bhawan, M-Block New Alipore, Kolkata – 700053.	For Identification of all species of animals.
Botanical Survey of India, CGO Complex, 3rd MSO Building, Block F (5th& 6th Floor), DF Block, Sector I Salt Lake City, Kolkata - 700 064	For Identification of plant species
Centre for Cellular & Molecular Biology Uppal Road, Habsiguda, Hyderabad - 500 007	For DNA profiling
Wood Properties and Uses Division Institute of Wood Science & Technology Malleswaram, Bangalore – 560 003	For Identification of Timber and wood properties.
Gujarat Forensic Science University, Near Police Bhawan, Sector 18 – A, Gandhi Nagar, Gujarat.	For Various capacity building courses on Forensic Science including WL Forensics.
State Forensic Science Laboratories (SFSL).	For forensic examinations of the samples in conventional crimes. These Laboratories can be approached for Wildlife Forensic Tests as well.

All Forest Officers especially the ones with biology background are qualified and competent enough to give such opinions, and their reports have same level of admissibility in courts as that of the scientists from WII or other institutes. For such identification, they may refer to the species identification protocols developed by the WII, Dehradun, and the trial courts should be explained/ convinced of it at the time of trial.

An effective search of a crime scene and surrounding area will provide important evidences.

1. Finger prints - The use of finger prints to identify individual is a commonly used forensic technique. Suitable surfaces at crime scene or on item potentially handled by a suspect may be appropriate for this method. Protect these from the effects of inclement weather prior to examination for finger prints.

2. Fibers, hairs and fur

3. Clothing or rope fibres snagged on tree bark during an egg collecting incident. Deer hair in the boot of poacher's car or on a knife.

4. Footwear, tyre and instrument marks. Footwear marks at a badger digging incident, climbing iron marks left on a tree by an egg collector, examination of close rings fitted to a birds which have been tampered with, Examination of quad bike tracks at a wildlife poisoning scene, Comparison of pliers seized from a suspect with the cut ends of razor wire put around a tree to protect rare resting birds

5. Soil, plant material (transferred to a spade during badger digging incident) and wood, paint (transfer or smashed headlight glass from vehicle used in a poaching incident), metal , plastics and petroleum based products, lubricants, adhesives, chemicals and building materials, glass samples associated with burglaries, vehicle glass left at road traffic accidents, bark and lichen on clothing and tree climbing irons used during an egg collecting incident

6. The study of pollen, plant spores, fungal spores and microscopic remains of plants and animals is called palynology. This study is useful for demonstrating contact between people, objects and places as well as for describing unknown places that have been contacted by a suspect, an animal or objects. The composition o. pollen grains at a crime scene is compared with samples taken from a suspect

clothing, vehicles, animals, tools etc. Even after washing the clothes the pollen can remain in pockets or cuffs. The potential source of pollen are Suspects - hair, hands, under nails, Clothing, shoes, fabric and plastic surfaces, digging implements, ropes etc, vehicle foot-wells, tyres, carpets, undersides, dirt, mud or dust recovered from a person or object, hair, fur, feathers, imported and exported goods(to verify the country of origin), packing materials particularly straw or cardboard, Animal remains such as soft tissues of stomach and intestines.

7. Obtaining palynomorph samples can be time consuming and they need to be examined by a competent palynologist., using a high power microscopy—light and phase contrast. A skilled palynologist can know the kind of place from which the pollen has been derived. The time of year can be determined as pollen is only released during certain season.

8. Feathers - Identification of feathers may be possible by looking at size, shape and colouration and making a comparison with reference collection held at museum. Licensed bird ringers are highly experienced in the identification of feathers from native species. Identification to species level may be possible by using DNA techniques.

Appendices

Appendix I   Guide to the Age of Animals

Teeth	Horse		Cattle	
	Temporary	Permanent	Temporary	Permanent
Incisors 1	Birth	2½ years		1½ years
2	1 month	3½ Years	Birth–	2+Years
3	9 months	4½ Years	One month	<3 Years
4				3+years
Canines		4years (Males)		
Pre-molars	Birth	2½ Years	Birth - one month	2 Years
		2½ Years		2+years
		3½Years		<3 Years
Molars		<1 Year		6 Months
		<2 Years		1+Years
		4 Years		2 Years

Teeth	Sheep		Dog	
	Temporary	Permanent	Temporary	Permanent
Incisors 1		1+years	3-4 Weeks	4-6 Months
2	Birth–	<2years		
3	One month	2+Years		
4		<3Years		
Canines			3-4 Weeks	4-6 Months
Pre-molars	Birth– One Month	<2Years	3 To 5 weeks	4 To 6 Months
		<2Years		
		2 Years		
Molars		3Months		6 To 9 Months
		9Months		
		1½Years		

Appendices

Horse (lower Incisors)

5 Years-13 apposition at anterior edges only
6 Years 11-180°
       I3- in wear
7 Years I1-Infundibulum disappears
       I3-(upper) -7 Years hook
8 Years I2-Infundibulum- Disappears
9 Years I3-Infundibulum –Disappears
10 Tears 13 (Upper) –Galvayne's Groove appears
15 years I3 (Upper)-Galvayne's Groove -1/2 way down
21 years I3 (upper) Galvayne's Groove-reaches tip
21 Years + I3(upper)-Galvane's groove-Gradually Grows out

Pig (Estimation of the age is not an accurate Method)

Teeth	Incisor1	Incisor2	Incisor3	Canines	PM-1	PM-2	PM-3	PM-4	M-1	M-2	M-3
Temp	2-4 Wk	Upper 2-3 Months Lower 1½ to 2 Months	Before birth	Before birth	5 Mon	5-7 Mon	Upper 4-8 Days	Upper 4-8 Days	4-6 Mon	8-12 Mon	18-20 Mon
Perm	12 Mon	16-20 Mon	8-10 Mon	8-10 Mon		12-15 Months	Lower 2-4 Wk	Lower 2-4 Wk			

Temp=Temporary; Perm=Permanent; Wk=Weeks; Mon=Months, PM=Premolars; M=Molars.

Determination of Age of Cattle

Cattle do not possess incisors on the upper jaw. Only dental pad is present. Cattle do not have canine teeth either. The incisors are arranged in horseshoe pattern and of chisel / shovel shaped. These incisors become straight with the advancement of age.

In the newly born calf, the umbilical cord will be dry and black by 4-5 days, falls off by 8-16 days, a cicatrix will form by 2-3 weeks and a scar disappears by 4 weeks. At birth or soon after all the incisors are present. If only 6 are present, the last pair will erupt in 2-6 days. Gums are red and almost cover the incisors. By 7-10 days gums retract. Traces of blue coloration will be visible on the gums in calves which are not more than 5 days old. By 15 days, the central and lateral incisors assume shovel formation. By 20 days, the corners assume free shovel shape. By 1 month all the incisors emerge from gums (pale pink colour).

In cattle above five years wearing of incisors on the labial surface begins. The wearing reaches half the level in the lingual surface around 7 years. Around 10 years wearing of incisors is seen in all the teeth in cingulum. In the cattle aged 11 to 12 years spacing of central incisors occurs and the roots of the teeth are exposed. In cattle aged around 12 to 13 years wearing of incisors where they were in contact, occurs.

Estimation of age of cattle by ossification of bones

Estimation of age with reasonable accuracy can be done by examination of carcass bones where the teeth are unavailable, based on the degree of ossification of certain parts of the cartilaginous extensions of the spines of the first five dorsal vertebrae.

Determination of age of cattle based on the ossification of cartilaginous extensions of first five dorsal vertebrae

Findings	Approximate age
Spines show entirely cartilaginous extensions, which are soft, pearly white and sharply delimited from bone that is soft and red.	1 year
Cartilaginous extensions show small red islets of bone	2 years
Numerous red / grayish areas of bone	3 years
Proportion of bone is greater than the cartilaginous areas	4-5 years
Complete ossification of cartilage with line of demarcation between the cartilage and bone still present	6 years

The changes are more rapid in cows and the ossification is complete after 3 years of age. In addition, other useful guides are (1) that the ischiopubic symphysis can be cut with a knife up to 3 years but not later when a saw is needed.

Appendices

In young bovines, the marrow in the vertebrae contains a red marrow but in the older ones the marrow is yellowish. The cartilage between the sternal segments is clearly visible but in the older animals it is bony uniformly.

Estimation of Age of Dog

Note: In dogs teeth cannot be used to determine the age due to difference in the size, rate of maturity and methods of feeding and development of teeth.

Estimation of Age of Camel

Approximate age	Characteristics
Approach of 5th year	Temporary incisors are replaced by permanent ones
Between 5.5 and 6 years	Permanent incisors, intermediate on either side of the central pair are replaced by fully developed ones
Between 6.5 and 7 years	Corner temporary teeth are replaced by permanent ones and incisors are complete
At 8 years	2 tushes in the lower jaw and 2 in the upper jaw
Between 8 and 11 years	6 tushes—2 anterior and 2 posterior in the upper jaw and 2 more in the lower jaw
At 11 years	Full mouth

Estimation of Age of Fetuses

The age of foetuses is estimated by measuring the crown-rump length.

Estimation of the Age of Bovine Foetus

Crown-rump length	Foetal Characteristics	Gestation in months
2 -5" or 6-7 Cm	Hooves discernible	2
4-6" or 10-17 Cm		3
10-12" or 25-30 Cm	Pigmentation and appearance of horn pits	4
12-16" or 30-40 Cm	Tactile hairs on upper eye lids and lips	5
20-24" or 50-60 Cm	Hair around eyes, horn pits, muzzle, and tip of tail	6
24-32" or 60-80 Cm	Fine hairs over the body and legs	7
	Complete hair coat	8
3 Feet long	Weight 35 Kg. The calf shows "golden slippers"	9
	Soft yellowish claws with convex sole	

Estimation of Age of an Equine Foetus

Crown-rump length	Foetal Characteristics	Gestation in months
$^3/_4$ " or 2 cm		1 month
2-3 " or 5-8 cm	Hooves discernible	2 months
3-6" or 8-15 cm		3 months
5-9" or 13-23 cm	Tactile hairs on upper and lower lips	4 months
8-14" or 20-35 cm	Tactile hairs on lips and upper eyelids	5 months
14-24" or 35-60 cm	Hairs on lips, nose, eye brows and eyelashes	6 months

Crown-rump length	Foetal Characteristics	Gestation in months
14-28" or 35-70 cm	Hairs on lips, nose, eyes and tips of tail	7 months
20-33" or 50-83 cm	Hair appears on mane and tails, ear, back and legs	8 months
24-26" or 60-90 cm	Hair (short and thin) on body except on abdomen	9 months
28-52" or 70-130 cm	Mane and tail well developed with hair coat complete but short	10 month
	Full hair coat	11 months

Estimation of Age of Foetus in Other Animals

Bitch

Gestation in days	10 days	10-21 days	3-4 weeks	5th week	6th week	7-8 weeks	9th week
Crown-rump length	Ovum 1/15, 1/20 inches	1/8 inch	1 inch	2 ½ inch	3 ½ inch	5 inches	6-8 inches (Kitten 5 inches)
Characteristics	Fertilized ovum in the uterus	Head, body and limbs discernible by the end of this period	First indication of hooves, claws visible (pale elevations at the end of digit	Well defined stomach	Large tactile hairs on lips, upper eye lids and above eye, teats visible in females	Well developed eye lashes, hairs on tail, head and extremities of limbs	Full size, Body fully covered with hairs, complete and soft

Sow

Gestation in days	14 days	3-4 weeks	4-6 weeks	6-8 weeks	8-10 weeks	11-14 weeks	15-17 weeks
Crown-rump length	Ovum 1/15. 1/20 inch	½ inch	1 3/4 inch	3 inches	5 inches	7 inches	9-10 inches
Characteristics	Fertilized ovum in the uterus	Head, body and limbs discernible by the end of this period	First indication of hooves, claws visible (pale elevations at the end of digit).	Well defined stomach	Large tactile hairs on lips,upper eye lids and above eye, teats visible in females	Well developed eye lashes, hairs on tail, head and extremities of limbs	Full size, Body fully covered with hairs, complete and soft

Doe

Gestation in days	14 days	3-4 weeks	5-7 weeks	7-9 weeks	10-13 weeks	13-18 weeks	18-21 weeks
Crown- rump length	Ovum 1/5, 1/20 inch	1/8 inch	$1^{1/4}$ inch	3½ inch	10 inches	1ft 2 inches	½ ft
Characteristics	--	Head, body and limbs discernible by the end of this period	First indication of hooves, claws visible (pale elevations at the end of digit	Well defined fore - stomach	Large tactile hairs on lips, upper eye lids and above eye, teats visible in females	Well developed eye lashes, hairs on tail, head and extremities of limbs	Full size, Body fully covered with hairs, complete and soft

Appendices 313

To know whether the calf was born alive or dead

In calf that was born dead, the lungs will be dark / pale grey in colour, firm to tough, dense in consistency and when a piece of the lungs is put in water it will sink indicating that the calf has not breathed.

Still birth: Expulsion of a dead foetus between 28-31 weeks of gestation.
Premature birth: Delivery of live calf before 28-31 weeks
Prenatal death or abortion: Expulsion of foetus prior to 25-40 days of gestation
Length of time lapsed since the birth of calf:- Based on foetal anatomical structures.

Determination of age of the newborn based on the characteristic findings of the foetal anatomical structures

Age in days	Characteristic findings
2-3 days	The umbilical artery closed completely. The umbilical vein and the ductus venosum starts closing
2-5 days	The umbilical vein and ductus venosum close completely.
3-4 days	The ductus arteriosus starts closing
7-10 days	The ductus arteriosus closes completely
Foetal anatomical structures	**Age in days**
If the umbilical artery is not closed fully but only narrowed	Calf has lived for 1 day
If the umbilical vein and ductus venosus are still open	calf has lived for 2 days
If umbilical artery, umbilical vein, ductus venosus and ductus arteriosus are closed but not Foramen Ovale	It indicates that the calf has lived for more than 10 days but less than 2 months.

Appendix II

Identification of sex of some of the wild animals / avian species.

At the time of inspection by regulatory authorities, the sex of the game carcasses is not grossly apparent.

1. Penile Ligament - In male artiodactylids (Order of even-toed mammals), portions of the penile ligament may be left behind in the field dressed carcass.

Differences in the pelvic bones in artiodactylids -Order of even – toed mammals

Structures	Males	Females
Suspensory tuberosities	present (serves for the attachment of the penile ligament)	absent
Ischial arch	V-shaped	U-shaped
Pubic arch	thicker	
Ventromedial border of acetabulum	thicker	
Position of the iliopectineal eminence (mature animals only)		

2. Cervical Vertebra - The sex of several species of artiodactylids may be determined by comparison of the cervical vertebra. To perform this technique, the soft tissue must be removed from the vertebrae and a transverse incision made through the column. Thus, this technique is more likely to be applied in the laboratory.

Structures	Males	Females
Accessory articular facets	Present	Not present

3. Lower Legs - Numerous measurements may be taken of the bones in the distal limbs of big game species to provide supportive information about the sex of an animal.

Structures	Males	Females
Measurements	larger than females	Thinner
Application of this data requires the use of regional standards.		

Appendices

Avian Species

(in field dressed waterfowl, pheasants, and turkeys).

Syrinx

The syrinx is the vocal structure of avian species. This structure may be left behind when the head is removed from a waterfowl carcass and can be palpated by inserting a finger through the thoracic inlet.

The syrinx is a sexually dimorphic structure in waterfowl. Under hormonal influence, the syrinx develops into a much larger structure in males.

Skeletal Measurements

Normally, the males are larger than females; however, significant overlap may occur. In mallard ducks, six measurements are used, including two sternal parameters, one scapular, one coracoidal, one pelvic, and one humeral

External Characteristics

Sexually dimorphic differences in plumage color and secondary sexual characteristics may aid investigators in determining sex based on partial carcass evidence, such as a pile of feathers or skin.

In some species, such as mallard and wood ducks, differences in plumage color are striking and can be differentiated easily between sexes. Keys exist to help distinguish feathers of sexes from species that initially may appear very similar.

Feather key for Ring-necked pheasants used to differentiate back feathers. Other species, such as turkeys, have pronounced secondary sex characteristics, such as beards and enlarged spurs.

Important anatomical points to remember

- Gannets do not have external nares but present inside mouth

- In several of the cranes, part of the trachea is coiled within the sternum.

- Uropygeal glands are very well developed in Spheniscidae, Podicipedidae (Grebes), Procellariidea, Laridae (gulls). Uropygeal glands are not present in Struthionidae, Rheidae, Casuariidae, Otididae (bustands), and may be absent or small in some Caprimulgidae, Columbiformes, Psittaciformes and Piciformes.

- Brood/incubation patches are thickened, highly vascular areas of featherless skin (one median or two lateral patches) in females and/or males depending on species These do not form in all species.

Heart: Pericardial sac is clear/translucent and contains a trace of clear fluid. Epicardial surface is clean and glistening The heart is contracted and triangular. Coronary groove separates the atria from the ventricles and usually contains fat deposits. No external demarcation between the left and right ventricles is visible. The atrioventricular valves are smooth and shiny. The left atrioventricular valve resembles that of mammals but the right atrioventricular valve is a thick muscular ring and not tricuspid.

Liver: Ostrich, many psittacines and most Columbidae lack a gall bladder.

Spleen: Oval and relatively large in pheasants. Normally small and flat in waterfowl There is no urinary bladder.

Testes: These are generally yellow in immature birds. In mature males in the breeding season these may be quite large and usually white with vascular surface The colour varies between species (sometimes melanistic).

Ovary: Paired ovaries are normal in some species. In Immature birds these are somewhat triangular, grey, with a rough-looking surface (immature follicles) and an inconspicuous narrow tubular oviduct extending to the cloaca. The oviduct in reproductively active females is a large tubular organ, off-white, flaccid, vascular and on the luminal surface, rugose.

The crop is a diverticulum of the oesophagus. Well developed in pigeons, gallinaceous girds and psittacines. In pigeons, it is divided into two lateral glandular sacs and produce crop "milk" for feeding chicks. In psittacines the crop is found transversely across the base of the neck.

Appendices

The gizzard in seed-eating birds has a thick muscular later and a hardened cuticle, the koilin or "keratinized layer". In penguins the stomach is distensible and reaches to the posterior abdomen.

Not all species have caecae, and some only have one.

The thyroids are paired, ovoid dark red or reddish brown.

The parathyroids are small and yellowish and may not be visible unless enlarged (dietary-induced secondary hyperparathyroidism).

Appendix III

Post mortem findings in some important pathological conditions in domestic animals.

1. Coma - Insensibility, resulting in death from different causes which in someway involve the central portion of the brain stem.

Symptoms

Sudden insensibility

- Loss of reflexes
- Relaxation of sphincters
- Dilated or contracted pupil and insensibility to light
- Cold skin. normal or subnormal temperature
- Slow, full and bounding pulse
- Slow and irregular breathing
- Death rattles (due to collection of mucus in the air passages)

Postmortem Findings

- Injuries to the skull bones or the brain.
- Effusion of the blood into the cranial cavity.
- Congestion of brain and meninges.
- Haemorrhages within the cranium due to diseases of meninges / brain.
- Clots of blood in between the skull bones and the membranes of the brain or on the surface of the brain if injury.
- Right side heart is full
- Left side heart empty
- Engorged venous systems and the lungs with blood but not so much as in death from asphyxia.

Appendices

Causes

1. Compression of the brain resulting from injuries or diseases of the brain or its membranes due to concussion, effusion of blood on or in the brain substance due to subarachnoid haemorrhage, fracture of the skull, inflammation , abscess or new growth in the brain, embolism or thrombosis.

2. Poisons, having a specific action on the brain and nervous system, such as opium, barbiturate, alcohol, carbolic acid etc.

3. Poisons that act on the brain after they are generated in the body in certain diseases of the liver and kidneys e.g. acetonaemia, uraemia etc.

2. Syncope - Death due to stoppage of the heart's action.

Symptoms

- Pale mucous membrane
- Dilated pupil
- Dim vision
- Great restlessness
- Air hunger
- Gasping for respiration
- Nausea possibly vomiting
- Marked fall in blood pressure (reflex vasodilatation)
- Slow weak pulse (fluttering in anaemia and rapid asthenia)
- Slight delirium, insensibility and convulsions preceding death
- In collapse, the animal retains consciousness (though the condition is attended with heart's failure).

Postmortem findings

- Heart in contracted state

- Empty chambers (when death occurs due to anaemia)

- Blood in both chambers in death due to asthenia.

- Pale lungs, brain and abdominal organs

Causes

1. Anaemia due to sudden and excessive haemorrhage from wounds of large blood vessels, or internal organs such as lungs, spleen etc or from bursting of an aneurysm or a varicose vein.

2. Shock due to sudden freight, blows on the heart or an epigastrium, drinking a large quantity of cold water when in a heated condition, extensive injuries to the spine or other parts of the body, sudden evacuation of natural or pathological fluids from the body or sudden pressure or severe exposure to cold.

3. Asthenia due to deficient power of the heart muscle as in fatty degeneration of the heart, aorta regurgitation and poisoning by certain poisons.

4. Exhausting diseases.

3. Asphyxia - Death due to stoppage of respiratory function (lack of oxygen) before the heart ceases to act.

Symptoms

Divided into 3 stages

Dyspnoea	Convulsions	Exhaustion
Animal shows anxious looks	Expiratory muscles of respiration are more active with spasmodic movements followed by convulsions of nearly all the muscles of the body	Paralysis of respiratory centre
Livid lips and prominent eyes	Deeply congested and cyanosed mucous membrane due to venous and capillary stagnation	Flaccid muscles, complete insensibility, loss of reflexes

Dyspnoea	Convulsions	Exhaustion
Deep, hurried and laboured respiration due to accumulation of carbon dioxide in the blood which stimulates the respiratory centre in the medulla.	Protruded tongue	Widely dilated pupil
Increased B.P. and rapid pulse	Loss of consciousness	Fall in B.P. Scarcelyp erceptible pulse but heart continues to beat for sometime after respiration has ceased. Prolonged sighings in respiration at longer and longer intervals until they altogether cease.
	Relaxed sphincters	Death

Postmortem appearances

External	Internal
Pale mucous membrane (slow asphyxia) or distorted, congested blue in sudden asphyxia	Mucous membrane of trachea and larynx cinnabar red in colour and contains froth
Protruded tongue	Dark and purple coloured lungs with engorged blood vessels filled with dark venous blood. On section frothy dark coloured fluid blood oozes out
Frothy and bloody mucus from mouth and nostrils	Distended or ruptured alveoli (emphysema)

External	Internal
Slow rigor mortis in some but may be rapid in some cases	Right ventricle, pulmonary artery and vena cava with dark coloured imperfectly clotted blood.
	Left ventricle, aorta and pulmonary veins empty. In many cases both sides of the heart full, if examined soon after death. But after rigor mortis has set in, the heart is contracted and the chambers are empty.
	Heavier lungs with blood collected in the dependent parts if examined sometime after death
	Congested brain (not as much as in death due to coma). Congested abdominal organs. Petechiae or ecchymoses over serous membranes(round, dark and well defined with varying in size-pin head to small lentil, over pleurae, pericardium, thymus, meninges of brain, cord, conjunctiva, epiglottis and under the skin of the face and neck. Similar haemorrhages are seen in scurvy, purpura or coronary thrombosis.

Causes

1. Mechanical obstruction of the air passages by foreign bodies, exudates, tumours, anaphylactic bronchospasm, suffocation, or drowning by blocking of their lumen from within, spasm of glottis due to mechanical irritation and irritant gases, forcible closure of mouth and nose by any means.

2. Absence of sufficient oxygen as at high altitudes or presence of inert gases, such as carbon monoxide in the atmosphere.

3. Stoppage of movements of chest due to exhaustion of respiratory muscles due to cold, debility, paralysis of respiratory muscles due to disease or injury to medulla or phrenic nerve or pneumogastric nerves, mechanical pressure on the chest or abdomen and tonic spasm due to tetanus or poisoning by strychnine.

Appendices 323

4. Collapse of lungs from penetrating wounds of the thorax or diseases such as pleurisy with effusion, empyaema or pneumothorax.

5. Pulmonary embolism- causing non entrance of blood into the lungs.

4. Classification of death - The cessation of the vital functions depends upon tissue anoxia.

Four ways of tissue anoxia

Anoxic anoxia	Anaemic anoxia	Histotoxic anoxia	Stagnant anoxia
Defective oxygenation of the blood in the lungs produced by 1. Obstruction to the passage of air into the respiratory tract as in suffocation, smothering and overlaying 2. Obstruction to the passage of air down the respiratory tract as in drowning, chocking from impaction of a foreign body, strangulation 3. External compression of the chest and abdominal walls e.g. from falls of earth 4. Primary cessation of respiratory movements causing respiratory failure e.g. from narcotic poisoning and from electric shock 5. Breathing in a vitiated atmosphere in which there is an excess CO_2 or inert gases.	In acute poisoning by Carbon monoxide, chlorates, nitrites and coal tar derivatives due to decreased oxygen carrying capacity of blood	Acute cyanide poisoning depression of oxidative processes in the tissues.	Insufficient circulation of blood through the tissues (stagnant anoxia), as occurring in deaths from traumatic shock, heatstroke and acute irritant and corrosive poisoning.

All these types of anoxia produce circulatory failure which may lead to death. In death due to sudden primary cardiac failure, there will be less visceral congestion than in the death occurring slowly.

4a. Sudden death

Causes

- It may occur due to unnatural causes such as violence or poison as well as from natural causes.

- Un-natural deaths have always to be investigated by the police. Very often natural deaths also form the basis of vetero-legal investigations if they occurred suddenly in apparently healthy animals and under suspicious circumstances.

- The possibility of death as a result of diseases and injury together has also to be kept in mind.

- Diseases of cardiovascular system causes sudden deaths. These are acute myocarditis as a result of infections like enteric fever. Diphtheria, pericarditis etc, fatty degeneration of the myocardium, rupture of a myocardial infarct or aneurysm.

- Left ventricular failure associated with aortic valvular disease, hyperthyroidism etc. Right ventricular failure associated with chronic emphysema and other lung diseases. Diseases of the pericardium and congenital abnormal conditions of the heart in calves.

Signs of death

1. Entire and permanent cessation of circulation and respiration
2. Changes in the eye
3. Changes in the skin
4. Cooling of the body
5. Cadaveric lividity-hypostasis, postmortem staining
6. Putrefaction or decomposition
7. Cadaveric changes in the muscle
8. Adipocere and
9. Mummification

Appendices

5. Suffocation - Death as a result of exclusion of air from the lungs by means other than compression of the neck.

Postmortem appearances

External	Internal
Death due to the cause producing suffocation or asphyxia	Mud or any other foreign matter in the mouth, throat, larynx or trachea if suffocation is caused by the impaction of a foreign substance in the air passages. These may be found in the pharynx and esophagus.
Suffocation: This is due to forcible application of the hand over the mouth and nostrils. Bruises and abrasions on the lips and angles of the mouth, along side of the nostrils. Lacerations on the inner side of the lips due to pressure on the teeth. Fracture of nasal septum due to pressure of hand (rare) Bruises and abrasions on the cheeks and molar regions or on the lower jaw, if there has been a struggle. Rarely fracture or dislocation of cervical vertebrae, if neck is forcibly wrenched in an attempt to smothering with hand. No local signs of violence if a soft cloth has been used to block the mouth and nostrils	Bright red mucosa of trachea covered with bloody froth.Congested and emphysematous lungs. Punctiform subpleural ecchymoses (Tardieu's spot) at the root, base and lower margins of lungs(not characteristic of suffocation since these also occur in death due to asphyxia from other causes. Similar haemorrhages in thymus, pericardium and along the roots of the coronary vessels. Lungs normal if death occurred rapidly. Right side of the heart is full with dark fluid blood and left side empty. Blood does not clot readily (so cuts after death show oozing of blood) Congested brain and abdominal organs especially liver, spleen and kidneys
Asphyxia: Open eyes, prominent eye balls and congested conjunctiva, livid lips and protruded tongue, bloody froth from mouth and nostrils.	

Causes

- Smothering or closure of the mouth and nostrils: Calves are accidentally smothered by being overlaid by their dams. The common method of killing the calves is to close the mouth and nostrils by means of hand or mud.

- Chocking or obstruction of the air passages from within: Mostly accidental and may be due to the presence of foreign bodies such as a piece of carrot, potato, corn, coin, cork, rag, round worm, mud, cotton and leaves. Even a small object blocking the lumen partially may cause death by spasm.

- Inhalation of irrespirable gases: Gases such Carbon dioxide (CO_2), an anaesthetic used by mistake, Carbon monoxide (CO), Hydrogen sulphide (H_2S) or smoke from a burning house will produce suffocation.

- Death is usually due to asphyxia or may be due to shock, when the heart stops by reflex action through the vagus nerves. Usually death occurs on a average from 10 to 15 minutes after complete withdrawal of air from the lungs,(fatal period) although instantaneous death occurs when wind pipe was blocked by a foreign body. Recovery may occur if treated within 5 minutes.

6. Drowning - Drowning is a form of death in which atmospheric air is prevented from entering the lungs by submersion of the body in water or in other fluid medium. Death is sure to occur even if the face alone is submerged so that the air is prevented from entering the respiratory orifices.

Rigor mortis may occur instantaneously and may pass off quickly.

1. Eyes half open or closed

2. Pupils dilated and conjunctiva congested

3. External injuries on the body due to struggling

4. Skin parchment like (sodden and wrinkled)

5. Water in rumen / stomach having similar characteristics as the water in which the animal was submerged

6. Fine froth around the nostrils and mouth. Large quantities of froth and fluid flow from nostrils.

Appendices

7. *Water in small intestines and middle ear

8. Fine white froth in larynx and trachea

9. Trachea and bronchi contain foreign bodies such as mud or fragments of water plants etc. Also clear or pink frothy foam. More froth in sea water drowning than in fresh water drowning.

10. On opening the thoracic cavity the lungs do not collapse but ballooned, heavier than normal, pits on pressure and show impressions of ribs. Dependent part of the lung lobes appear atelectic and haemorrhagic.

11. Large blood vessels are engorged

12. Liver congested

13. Spleen and kidney enlarged and congested

14. In males, the scrotum and the penis are congested.

15. Heart: Right ventricle and auricle dilated and filled with dark fluidy blood. Left side contracted and empty

* sure sign of drowning

Differences between ante-mortem drowning and drowning of a dead body

Ante-mortem drowning	Drowning of dead body
Fine froth around the nostrils and mouth	Not present
Injuries on the body due to struggling with inflammatory reaction	Injuries caused on the dead animal do not show inflammatory reaction
The distended body will float within a few hours and without putrefactive changes	The body will show early putrefactive changes
Water will be present in the middle ear and intestines	Not possible for the water to enter these parts after death

*Confirmed by finding diatoms by microscopic examination of liver and bone marrow (the diatoms reach liver and bone marrow). Due to the pumping action of the heart when the animal is struggling the diatoms reach the liver and bone marrow.(the diatoms are found in the lungs even in the dead carcasses thrown in the water but not in the liver and bone marrow).

Modes of Death

Asphyxia	Shock	Concussion	Apoplexy	Exhaustion
Most common cause in majority of cases, as the water getting into the lungs gets churned up with air and mucus and produce fine froth which blocks the air vesicles. Occasionally death occurs from asphyxia caused by laryngeal spam set up by a small amount of water entering the larynx. In such cases water does not enter the lungs and signs of drowning will be absent.	Due to fright or terror or it may be caused by sudden and unexpected fall in water. If water is very cold it may induce laryngeal shock resulting in cardiac arrest due to vagal inhibition through the sensory laryngeal nerve endings. Shock may also be induced through the sensory nerve endings of the cutaneous nerves.	Due to fall in the water with head or buttocks from a height and striking against some hard solid substance or even against water itself.	Cerebral vessels may rupture (if diseased) due to sudden rush of blood to the brain due to cold, excitement or the first violent struggle to keep above the surface of water.	Due to continued efforts to keep above the surface of the water.

The most characteristic sign is the presence of froth at nostrils and mouth. Distended lungs, paler than in other forms of asphyxia, some alveolar emphysema and bronchi and bronchioles full with fine froth.

7. Death from starvation / Inanition

Results due to deprivation of a regular and constant supply of food. Acute occurs when there is sudden and complete withholding of food and chronic when there is a gradual deficient supply of food.

Appendices

Symptoms

- In protracted absence of food: acute feeling of hunger lasts for the first 30-48 hrs, accompanied by intense thirst

- After 4 or 5 days, emaciation and absorption of subcutaneous fat begins

- Sunken and glistening eyes

- Widely dilated pupil

- Dry and cracked lips and tongue

- Ulcers on the gums with foetid smell. Dry mouth and tongue.

- Offensive and foul breath

- Faint and weak voice

- Dry, rough wrinkled and baggy skin with peculiar and disagreeable odour

- Weak and slow pulse which accelerates on slightest exercise

- Subnormal body temperature

- Sunken abdomen and loss of muscular power

- First there is constipation but followed by diarrhea or dysentery towards death

- Scanty , turbid and highly concentrated (acidic) urine

- Loss of weight marked and constant (Loss of 40% of the body weight leads to death).

- Occasionally delirium and convulsions or coma precede death.

- Death occurs in 12-15 days (Fatal period) if both water and food are totally deprived. If food alone is deprived life may be prolonged for a long period (6-8 weeks or more).

Postmortem appearance

Rigor mortis commences early.

External	Internal
Greatly emaciated carcass Dry eyes, red and open and sunken eye ball	Brain normal, but sometimes pale and soft Congested meningeal vessels
Dry and coated tongue Dry and shriveled skin, hide- bound with dry, brittle and lusterless hairs Bed sores if lying down Pale, soft and wasted muscle Absence of subcutaneous fat Loss of fat in omentum, mesentery and internal organs Entire absence of fat throughout the body is never seen in wasting diseases (As in TB and JD). Fragile long bones	Serous effusions in ventricles Heart small in size, muscle pale and flabby and chambers empty. Pale, collapsed lungs with little blood when cut. At times edematous and show hypostatic congestion at the bases. Stomach reduced, contracted and empty. Undigested food if given before death. Mucous membrane of stomach and upper part of small intestines is more or less stained with bile. Contracted and empty intestines. Lower portions of large intestines contain hard faecal matter and may show evidence of inflammation. Extremely thin and translucent intestinal wall indicating that no food passed through the gastrointestinal tract for considerable time. Sometimes ulcers may be seen. Viscera-All organs reduced in size. Small and shrunken liver, spleen, kidneys and pancreas Distended gall bladder with dark inspissated bile Empty urinary bladder.

Progressive wasting and emaciation is seen in TB, malignant disease, progressive muscular atrophy, pernicious anaemia and chronic diarrhea. One must examine the internal organs of the animal very carefully and search for the existence of these diseases before giving opinion that the death is due to starvation. Check for malignant disease, chronic disease, tuberculosis, and pernicious anaemia

Appendices   331

8. Burns

- Injuries produced by flame, radiant heat or some heated solid substance like metal or glass. Injuries caused by friction, lightning, electricity, X rays and corrosive chemicals are all classified as burns for vetero-legal purposes.

- Scalds are moist heat injuries. Injuries caused by the contact of the body with a steam liquid at or near its boiling point or in its gaseous form. These are not so severe as burns. These cause hyperaemia and vesiculation. The lesions produced by heated oil resemble burns very much in severity. Scalds produced by molten metals cause great destruction of the tissues as they strongly adhere to the parts struck

- Burns due to X rays occur due to prolonged exposure. The area appears from redness of skin to dermatitis with shedding of the hair and epidermis and pigmentation of the surrounding skin. Severe exposure produce vesicles or pustules which often form ulcers and take long time to heal.

- Burns caused by radium are very similar to X ray burns.

- Prolonged exposure to sun also causes burns. The ultraviolet rays may produce erythema or acute eczematous dermatitis of the exposed part.

- Burns produced by corrosive chemical (strong acids and caustic alkalis) result in eschars (soft, moist and readily slough away). There is no line of demarcation, hairs are not scorched nor are vesicles formed. The characteristic stain found on the skin assist in determining the nature of the corrosive used.

- Burns do not result in death but the injuries may be serious resulting in loss of sight or permanent disfigurement.

Classification

1st degree	2nd Degree	3rd Degree	4th Degree	5th Degree	6th Degree
Erythema or simple redness of skin due to momentary contact with flame or hot solids or liquids much below boiling point. Also produced by mild irritants. The redness and swelling disappear in a few hours but lasts for several days if upper layer of the skin peels off. No scar results since there is no destruction of tissue.	Involves acute inflammation and formation of vesicles due to prolonged contact with flame, liquids at boiling point or solids. Vesicles are produced by strong irritants and vesicants such as cantharides. In burns caused by flame or a heated solid substance, skin is blackened and hair singed at the seat of lesion. No scar since only superficial epithelium is destroyed.	Involves the destruction of cuticle and part of the true skin which appears horny and dark due to charring and shriveling. It is painful since nerve endings are exposed. This leaves scar but there is no contraction since the scar contains all the elements of true skin.	There is destruction of whole skin. The sloughs are yellowish brown and parchment like and separates out from skin by 6 to 7 days leaving an ulcer which heals slowly forming scar of dense fibrous tissue with consequent contraction. Burns are not painful due to complete destruction of nerve endings	Involves deep fascia and muscles. Result in great scarring and deformity.	Involves charring of whole limb and ends in inflammation of adjacent tissues and organs if death is not the immediate result.

Appendices

Effects

Burns and scalds vary in severity depending on

- Degree of heat: Very great heat produces more severe effects

- Duration of exposure: Symptoms are more severe if heat is applied continuously for long time

- Extent of surface: If one third to one half of the superficial surface of the body is involved it ends fatally

- Site: Extensive burns of thoracic and abdominal regions, even though superficial are more dangerous than those of extremities. Burns of genital organs and lower parts of abdomen are often fatal)

Causes of Death

1. Shock	2. Suffocation	3. Accidents / Injuries	4. Others
Severe pain due to extensive burns causes shock to the nervous system and result in a feeble pulse, pale and cold skin and collapse, leading to death instantaneously or with in 24–48 hrs Shock may be due to fright (if the heart is weak or diseased) before the individual is affected by burns. If death does not occur due to shock, it may occur subsequently due to toxaemia (absorption of toxic products from the injured tissues in the burned area). The temperature rises to 106°F, pulse increases and the animal becomes restless and passes into unconscious state and death supervenes due to prolonged state of shock.	Animals removed from houses destroyed by fire are found dead from suffocation due to inhalation of smoke, Carbon dioxide (CO2) and Carbon monoxide (CO), products of combustion. In such cases burns occur after death.	Occur while escaping from the burned area or by falling walls or timber	Inflammation of serous membranes and internal organs such as meningitis, peritonitis, edema, glossitis, pleurisy, bronchopneumonia, pneumonia, enteritis and perforating ulcer of the duodenum, Hypoproteinaemia and anaemia Exhaustion from suppurative discharges lasting for weeks or months Septicaemia, pyaemia, gangrene and tetanus.

Death may occur within 24 to 48 hrs. Most fatalities occur in the first week. In suppurative cases it occurs after 5 or 6 weeks or even longer.

Postmortem appearances

External	Internal
Remove the carcass carefully and examine for the presence of kerosene, petrol or any other combustible or inflammable substance. External appearance varies with the nature of substance	Congested brain and its meninges Extravasation of blood in the Dura mater Brick red or reddish brown deposit upon the surface of the Dura mater Shrunken brain (form is retained).
Radiant heat—Skin is whitened	If death is due to suffocation, the larynx, trachea and bronchi contain soot with congested mucous membrane and covered with frothy mucus (absence of soot indicates that the animal was not alive at the time of fire).
Burns by flame-May or may not produce vesication, but singing of hair and eye brows and blackening of skin are always present. With a red hot body or a molten metal—On contact, blister and reddening corresponding in size and shape to hot material occur but cause roasting and charring of the body parts when in contact for a long time.	Congested and inflamed pleura with serous effusion in the cavity.
Burns caused by kerosene are very severe and identified by odour and sooty blackening of the parts of the body	Congested and shrunken lungs, rarely anaemic Heart chambers full with blood Cherry red coloured blood if CO is inhaled produced by incomplete combustion.
When the body is exposed to great heat-tissues get cooked and becomes rigid with flexing of limbs (due to coagulation of its albuminous constituents). With very great heat—cracks and fissures resembling incised wound occur in the skin and tissues, but no blood clot nor infiltration of blood is found in the cellular spaces and blood vessels are seen stretched across the fissures, since they are not usually burnt. The charred skin (hard and brittle) cracks easily when attempted to remove the body from a place destroyed by fire. Scalds caused by boiling water or steam produce reddening and vesication but do not affect the hair and do not blacken or char the skin. Super heated steam soddens the skin and becomes dirty white.	Red mucous membrane of stomach and intestines Inflammation and ulceration of payer's patches and solitary glands of intestines. Ulcers in the duodenum if death occurs sometime after burns. Enlarged and softened spleen Cloudy swelling and necrosis in liver. Kidneys shows signs of nephritis and the straight tubules contain debris of red blood corpuscles (reddish brown markings) Enlarged and congested adrenals.

Appendices

Difference between ante-mortem and postmortem burns

Three points of difference between ante-mortem and postmortem burns relate to

	Ante-mortem burns	Post-mortem burns
Line of redness	A line of redness involves whole thickness of skin around the injured part. It is a permanent line and persists even after death. Erythema due to distension of capillaries found beyond this line of redness is transient disappears under pressure during life and fades after death. The line of redness separates living from the dead tissues and is always present, though it takes sometime to appear.	Not seen.
Vesication	Contains serous fluid consisting of albumin and chlorides and has a red inflamed base with raised papillae. Skin surrounding it is bright red or copper coloured (True vesication) in contrast to the false vesication which is produced after death.	Vesicles contain air only but may contain a very small quantity of serum comprising trace of albumin but not chlorides. Base is hard, dry and yellow
Reparative process	Signs of inflammation, formation of granulation tissue, pus and sloughs are noticed - indicate that the burns were caused during life.	Dull white appearance and the openings of the skin glands are grey in colour. Roasted internal organs which emit a peculiar offensive odour.

Time of burns

Appearance	Time
Redness	Immediately after burn
Vesicles	2 to 3 hours
Formation of pus	2 to 3 days but not before 36 hours
Separation of Superficial sloughs	4th to 6th day
Separation of deep sloughs	Within a fortnight
Covering of surface of burns with granulation tissue	Beyond fortnight
Cicatrix and deformity	After several weeks / months depending on the amount of suppuration, sloughing and depth and extent of burns.

9. Frost bite

1. Rigor mortis appears slow and lasts longer

2. Vesicles on the skin (Escharotics- dry/moist necrotic mass of tissue)

3. Patches of pink discolouration on the lateral aspect of hip, elbow and knee joints as well as on the flank and face

4. Edematous swelling of the ear, nose and toes.

5. Heart: Full with bright red clotted blood

6. Superficial erosions in rumen and abomasum/ stomach

Confirmation: Based on circumstantial evidence- season, atmospheric temperature, absence of injuries, place where the animal was found etc

10. Lightning

- Lightning is the discharge of atmospheric electricity in the open field or in the shed, especially if the animals are standing near the open doors and windows through which the discharge enters, during a thunderstorm

- Lightning is attracted by tall objects like tall trees during thunder storms and hence it is dangerous to keep the animals below tall trees.

Appendices

- When the animal is struck by lightning, it falls unconscious due to syncope or concussion and dies at once from its paralyzing effect on the nervous system or subsequently from the effects of burns and lacerations after some days or even weeks.

Postmortem appearances

- Appearance is varied.

- Ecchymoses, contusions, lacerations, wounds of any type, simple, compound and comminuted fractures of bones and burns caused by superheated air varying in depth and extent.

- Singing of hair, blisters, fissures and charring caused by burns

- Skin burns--Reddish brown arborescent markings on the surface of skin resembling the branches of a tree- characteristic of lightning, singing of hairs, blisters, charring and fissures occur. These markings are superficial burns caused by super heated air which produce mere erythema of the skin indicate the path taken by the branching nature of discharge).

- Animal dies instantaneously (due to electrical discharge passing to earth).

- Injuries are also caused by blast caused by forceful displacement of air around the lightning flash followed by compression due to forceful return of air.

- The metallic articles carried by animal are fused and steel articles are magnetized which may leave their impressions on the skin.

Before giving opinion as cause of death is due to lightning, the following points should be remembered.

- History of thunder-storm in the locality

- Evidence of effects of lightning in the vicinity e.g. damage to the houses or trees, death of men and animals

- Fusion or magnetization of metallic substances

- Absence of wounds and other injuries.

Postmortem appearance

External	Internal
1.Rigor mortis sets in soon after death and pass of quickly. Above lesions are usually present on the body of the animal. 2.Congested eyes and dilated pupil 3.Superficial lymph nodes haemorrhagic 4.Blood stained froth from nostrils and mouth. Anus relaxed	Not characteristic. Extensive haemorrhages and lacerations in the brain. Petechial haemorrhages on the pericardium. Heart- left side empty and right side full. Petechial haemorrhages on the mucous membranes of internal organs, serous membranes, trachea, endocardium, meninges and brain, Lungs show petechiae and congestion. Blood fluidy, tarry and unclotted Rupture of blood vessels and tearing of internal organs Congested viscera

11. Electricity - Death from electric currents is mostly accidental.

Effects

Chief effect is shock and varies with

Nature of current	Resistance of the body
Current generated at high voltage are dangerous to life. Alternating currents are considered more dangerous than the direct currents because they are generated at high tensions. Direct current at high tension is equally dangerous. Alternating currents of low frequency are dangerous even at low voltage but danger diminishes with the increase of frequency even when generated at high voltages. Besides high voltage, close contact of long duration increase the danger.	The effect of electric shock varies with the amount of resistance offered to the flow of current. Amperage is more important than voltage Animal body is a bad conductor of electricity though resistance of different tissues varies. Skin offers greatest resistance. Oily and hard skins are more resistant than moist, soft and perspiring skins. Resistance diminishes with the continuance of the current. Also diminishes in cardiac disease, kidney disease and surprised shock.

Appendices

Symptoms

Burns at the point of entrance and exit of an electric current (at these points the skin offers resistance and the electrical energy is changed into varying degrees of intense heat). Sparking causes charring of tissues, suffused eyes, dilated pupils, cold and clammy skin, stertorous breathing and insensibility. In severe cases, insensibility may occur immediately and may be followed by death.

Cause of death

Due to sudden stoppage of the action of the heart and inhibition of respiratory centre in the brain stem.

Fibrillation of cardiac ventricles is more common after contact with low voltage circuits while at high voltage death is due to respiratory failure due to central inhibition of the nervous system.

Sometimes death may occur later due to complication of electrical injury like burns and infection.

Postmortem appearance

External	Internal
Congested eyes	Congested internal organs
Dilated pupil	Edematous lungs
Lesions at the point of entry and exit of electric current	Petechial haemorrhages in meninges and pleura, pericardium and endocardium
Microscopically: at the site of an electrical lesion	Ecchymoses along the path of current.
Compression of horny layer into a homogeneous plaque	
Fissures and hollows between the corneum and germinativum	
Coalescence into a star shaped or rod like structure of the basal cells in each group of the rete Malpighi—surest sign that the electric current has passed.	

12. Snake bite

1. Local - blood stained, edematous swelling
2. Fang marks when skin reflected
3. Passive congestion of organs

13. Blast injuries in animals - Salient features.

Primary Category - Causes direct effects. The pressure wave damages external and internal tissues which are often fatal. Burns may occur.

Secondary Category - Caused by wood splinters, broken glasses etc. The flying objects cause external damage.

Tertiary category - The body of the animal is thrown about (especially the small and light species such as birds). There will be external and internal damage and concussion.

14. Ruminal tympany

- Over distension of the rumen and reticulum with gases

- Protrusion and congestion of tongue

- Marked congestion and haemorrhage of lymph nodes of the head and neck, epicardium and upper respiratory tract

- Friable kidneys and hyperaemia of intestinal mucosa

- Compressed lungs,

- Congestion and haemorrhage of cervical esophagus but thoracic esophagus is pale and blanched

- Distended rumen but the contents are much less frothy than before death

- Pale liver due to expulsion of blood from the organs

- In animals that were dead for some hours--Subcutaneous emphysema, complete absence of froth in the rumen, exfoliation of the cornified epithelium of the rumen with marked congestion of sub-mucosal tissue.

Appendices

15. Anthrax

- No rigor mortis

- Rapid gaseous decomposition

- Saw horse appearance

- Dark tarry blood from natural orifices that does not clot

- Rapid putrefaction and bloating

- Ecchymotic haemorrhages throughout the body

- Blood stained serous fluid in the body cavities

- Severe enteritis

- Enlarged spleen with softening and liquefaction of its structure.

Appendix IV

Pathological Conditions based on Gross Appearance of Carcass

Type of disease suspected based on the odour emitted from the body

Nature of odour	Disease suspected
Ammonical smell	Uraemia
Bitter almond odour	Sweet clover poisoning
Garlic odour	Phosphorus poisoning
Peculiar sweet odour of acetone	Ketosis
On opening stomach-Acetylene odour(smell between garlic and rotten fish)	Zinc Phosphate poisoning

Type of disease suspected based on the condition of the skin

Condition of the skin	Disease suspected
1a. Yellowish discolouration	Anaemia, Jaundice, Nitric acid poisoning
b. Very faint cherry pink discolouration of mucous membrane	Carbon monoxide poisoning (may be missed)
c. Cherry pink or red discolouration of mucous membrane	Cyanide poisoning and methaemoglobinaemia due to several chemical poisoning Viz. Phenacetin and some nitro compounds, burns, Hydrocyanic acid poisoning
d. Intense bluish violet or purple colouration of mucous membrane / bright scarlet red skin	Asphyxia
e. Chocolate or coffee brown colouration of mucous membrane	Potassium chlorate, Potassium dichromate or Aniline poisoning
f. Leathery skin, which cracks and peels leaving raw areas	Arsenic poisoning

Appendices

Condition of the skin	Disease suspected
2. Erythematous rashes (flushed skin)	Datura poisoning
3. Extensive ulceration, sloughing of the skin over muzzle, near lips, and perineal region	Lantana poisoning
4. Edematous swelling of throat and dewlap	Mercury poisoning
5. Scaly skin – parakeratosis	Zinc deficiency
6. Dull and copper coloured hairs in black coated animals	Cobalt and Zinc deficiency
7. Dry skin with broad wrinkling without elasticity and pigment / edema and ulcers at different places on the body	Starvation
8. Skin wet, wrinkled and parchment like	Drowning
9. Third degree burn marks at the point of entry	Electricity marks
10. Petechial haemorrhages in the skin over the neck thorax, loin and back	Violent death
11. Arborescent markings- tree shaped branching, reddish or reddish blue streaks (due to super heated air) with singing of hairs, blisters, charring and fissures (resembling tree), lacerated wounds and contusions	Lightning
12. Bright cherry red skin and dry and lustreless hair	Frostbite

Type of disease suspected based on the appearance of the visible mucous membrane

Condition of the mucous membrane	Disease suspected
1a. Anaemic / Pale b. Blanched	Sweet clover poisoning, Lead poisoning
2. Cyanosed-dark reddish blue	Chloral hydras and cyanide poisoning
3a. Markedly Icteric b. Slight orange to intense yellow with greenish tint	Chronic wasting disease Jaundice

4a. Petechial haemorrhages
  b. Larger haemorrhages
  c. Icteric with petechiae
  d. Congested
  e. Muddy appearance
	Carbon tetrachloride poisoning
	Trauma, inflammation,
	haemorrhagic diathesis
	Surra, Theileriasis
	Datura poisoning,
	Methaemoglobinaemia

5a. Congestion of conjunctiva
  b. Pin point haemorrhages over conjunctiva
  c. Dirty brown conjunctiva
	Colic
	Purpura, drowning
	Nitrite poisoning

Collect tissues from all the organs (1 to 2 cm cube) and fix in 10% formal saline (10 times the volume of specimens) for histopathological and special examination.

Appendices 345

Subcutaneous Tissue

Type of disease suspected based on the findings in the subcutaneous tissue

Findings	Disease suspected
1. Petechial or ecchymotic haemorrhages in the subcutaneous tissue, muscles of leg, breast and body cavities	Dicoumarin and warfarin poisoning (Sweet clover), Bracken fern poisoning
2. Extravasation of blood in subcutaneous tissue and muscle	Phosphorus poisoning
3. Dislocation, fracture of bones and haemorrhages in subcutaneous tissue (s/c) and muscle	Crushing injuries
4. Generalized subcutaneous edema	Heart failure
5. Pale muscles and subcutaneous tissues	Internal bleeding, mercury poisoning
6. Haemorrhages in the muscle	Sweet clover poisoning and vitamin K deficiency

Time of Hypostasis

Within about one hour after death, red patches coalesce with each other forming extensive areas of reddish-purple discoloration	Visible two to three hours after death
Fixed in its primary position	After 6 hours.
Complete and maximum intensity	After 10 hours.

Position of Hypostasis

Primary position	Lower side of the dead body
In Drowning	Appears in the head, neck and upper parts of the chest
In hanging	Appears in the legs and the lower parts of the dead body.

Extent of Hypostasis

Minimal	In deaths from hemorrhage due to the loss of blood volume.
Less marked	In malnutrition and cachectic animals.
Extensive	In congestive heart failure and long-standing diseases

Difference between hypostatic congestion and postmortem congestion

Hypostatic congestion	Post mortem congestion
The bronchi and trachea contain fine froth	No such froth noticed
Mucous membrane is bright pink, soft and elastic	Dull and lusterless
Exudate may be noticed	Not so
Pleura may be covered with fibrinous exudates	Not so

Similarly the postmortem staining of tissues should be differentiated from the bruise.

Differences between postmortem staining and bruise

Postmortem staining/ lividity	Bruise
This occurs over an extensive area of the most dependent parts of the body and usually involves superficial layers of the skin	This occurs anywhere on the body. Usually takes the shape of a weapon. It is limited in area and usually affects deeper tissues
It does not appear elevated above the surface and has sharply defined edges	This appears raised and has no sharp edge
The colour is uniform and becomes green when the body begins to putrefy.	This exhibits usual colour changes
No abrasions are noticed	Abrasions are noticed
Does not show any effusion when cut	Shows infiltration of tissues with coagulated or liquid blood
Postmortem staining/ lividity	Congestion
In the internal organ it is irregular and occurs in the dependent parts.	The redness caused is uniform and seen all over the organ
The mucous membrane is dull and lusterless	The mucous membrane is moist and glistening

Appendices

Colour changes - age of bruise

Determination of the age of the bruise based on colour changes

Red	1st day	Greenish yellow	5-6 days
Blue	2nd day	Yellow	7-12 days
Bluish black brown / livid red	3 days	Normal colour	14-15 days (Yellow fades)

Disappearance is more rapid in healthy animals
Ante mortem bruise: Swelling, colour changes and blood clot in s/c tissue
Post mortem bruise: Swelling- absent, blood clot in s/c tissue and muscle. If bruise is made soon after death it is difficult to distinguish from ante mortem bruise.

Estimation of the Age of Injury

The age of bruise may be ascertained from the colour changes which its ecchymosis undergoes. These changes start from 18 to 24 hours after its infliction.

3-5 hours after injury---polymorphs in a haematoma begin to disintegrate. They are fragmented within 21 hours and by 30 hours the basophilic nuclear fragments have either undergone autolysis or have been engulfed by monocytes.

The age of the wound may be ascertained by observing the following stages of healing process

The wound (aseptic) is covered with lymph in 36-48 hours'
Edges join together by 72 hours
Wound heals by 1st intension by 7th to 8th day-a red tender linear scar appears. In vascular parts heals by 4-6 days
Septic wound with loss of tissue heals by formation of granulation tissue.
Edges are bound together by blood and lymph during first 12 hours. Margins are red and slightly swollen with leukocytic infiltration
By 24 hours proliferation of vascular endothelium occurs
36 to 40 hours new capillary vessels are formed

48 to 56hours spindle shaped cells run at right angles to the vessels in the deeper parts of the wound visible

3- 6 days definite fibrils of connective tissue run parallel to the long fibroblasts

3-4 weeks dense fibrous tissue formed

In septic wound pus appears in 36 to 48 hours.

In Bone

1-3 days—Signs of inflammation and exudation of blood in the soft parts and around fractured ends

4-15 days soft tissue callus is noticed

15 days to 5th week- ossification occurs

6 to 8th week—union of fractured end with formation of bone

In the skull bones

Within a week—very slight amount of callus formation occurs and the edges of a fissured fracture are glued together

Within 3-4 weeks—gradually become smooth

2-3 months or more—formation of bone occurs.

In comminuted fractures the line of fracture remains permanently visible. Gaps are filled with fibrous tissue.

In case of dislocation of joint the time can be judged from the colour changes of the bruise

If tooth is lost, bleeding from its socket may continue for a few hours and stops in about 24 hours.

In 7-10 days the cavity fills up.

After 14 days- Alveolar process becomes quite smooth.

Differences between incision and laceration

Feature	Incision	Laceration
Appearance	Sharply outlined, linear, curved, or angular. Bruising is rare. Strands of tissue bridging the wound rarely present.	Often irregularly outlined, round or oval or stellate, Bruising is common, Tissue bridges often present.
Pelage	Little or no loss of hair or feathers around wound.	Hairs or feathers may be absent.
Haemorrhage	Marked	Minimal or apparently absent

Other Types of Injuries

- Rupture - It is an injury in which the tissues are stretched until the fibres part. It is caused by a severe crushing blow or extensive distension. It may involve the wall of hollow organ (urinary bladder or gall bladder) or capsule of parenchymatous organ (liver and spleen) or tissues (Muscles, tendons, ligaments).

- Fracture - It is an injury of bone, cartilage, tooth, hoof, horn or claw in which the continuity of hard structure is broken. If the skin is not broken, it is called a simple fracture and if the integument is broken it is known as compound fracture. The cutaneous defect enables bacteria to enter the area, the resulting inflammation interferes with bone healing.

- Concussion - A term used to describe a functional disturbance of central nervous system, which may or may not be associated with loss of consciousness following a severe jarring injury to the head. In this, the cranial vault may or may not be broken. Haemorrhage may or may not occur. The jarring initiates pressure waves that rebound back and forth through the semi fluid nervous tissue resulting in unconsciousness or in-coordinated movements of the animal.

- Sprain or Strain - An injury of a joint in which the anatomic relations of the bones are maintained but the supporting ligaments around the joint have been stretched or torn slightly.

- Luxation or the dislocation - An injury of a joint in which the anatomic relations of the bony structures are not maintained and ligaments supporting the joint are torn.

Points to remember on wound and wound healing in animals.

- The veterinarian's knowledge of animal behavior, clinical experience, and common sense must be used to evaluate injuries.

- There are several considerations to make this determination depending on the type of injury.

- Wound healing in cats generally occurs more slowly than in dogs.

- In ponies, the early inflammatory response is more intense than in the horse, leading to uncomplicated healing. The second intention healing in ponies is significantly quicker compared with horses, with the rate of wound contraction being better.

- Responses to cutaneous wounding differ in horses and cattle, with horses showing a more rapid development of granulation tissue, while growth and differentiation of connective tissue in cattle 10 days post injury is more pronounced than in horses.

- In deceased victims, the first thing to determine is the fatal injury and then back track from there.

- For contusions and areas of hemorrhage, one must remember that hemorrhage requires a beating heart. In multiple stab wounds, there may be little to no hemorrhage around an injury that was made when or after the heart stopped beating. But, if there is minimal blood supply to the area then it could account for the minimal amount of hemorrhage.

- With multiple fractures or injuries, a diagram can help in the evaluation of the number of blows and determining sequence of events. With multiple fractures of the skull, one blow may cause concentric or radiating fractures. Evaluation of where these lines stop and start can help determine the number and type of impacts. Consideration must be given to the impact each injury would have had to the animal. This includes how the injury would have compromised the animal such as severe pain, the ability to move, vocalize, or fight back. In addition, the veterinarian must consider what the animal's response would have been to each injury.

- This is critical in courtroom where the veterinarian must testify about the expected reaction and vocalization to each event/injury as part of their expert witness testimony.

Appendices

Type of disease suspected based on the type of effusion in the serous cavities

Serous effusions	J.D. Struck (peritoneum- with clots), Ephemeral fever, Anaplasmosis, Heart water, Theileriasis, Trypanosomiasis and Liver fluke infestation
Serosanguineous effusions	Malignant edema, Braxy (pericardium-large amounts of clots), Bacillary haemoglobinuria, Lamb dysentery (peritoneum), Enterotoxaemia (pericardial sac), Haemorrhagic septicaemia (peritoneum), Canine distemper (pericardium), Infectious hepatitis (peritoneum), Blue tongue (pericardium) and Babesiosis.
Sero-fibrinous exudates	Erysipelas, Haemorrhagic septicaemia (pleura), Contagious bovine pleuropneumonia (pleura), Contagious caprine pleuropneumonia (pleura) Brucellosis and Canine distemper (pleura)
Clusters of/ bunch of tubercles- Pearls, thickened pleura / peritoneum	Tuberculosis

Appendix V

Determination of time of death based on the presence of rigor mortis on the body (Domestic Livestock)

Condition of the body	Approximate time of death
A. If the dead body is very fresh with no postmortem changes and with clotted blood of recent origin	Indicates that death has taken place 2-4 hrs earlier from the time of examination
B. If postmortem changes are seen	The time of death is determined on (from the time of examination) the basis of rigormortis and the status of dead body.

	Winter	Summer
1. Rigor mortis appears in the head	Within 2-8 hrs	Within 30 mt to 3 hrs
2. Rigor mortis is present in head, neck and fore limbs	12 hrs	
3. Rigor mortis is present in whole body	15 hrs (15-20 hrs)	
4. If rigor mortis is present in hind limbs alone	20 hrs	
5. If rigor mortis disappeared from whole body	24-30 hrs	rs
6. Body loose, bloated with putrid or obnoxious smell.	More than 30hrs (Putrefaction)	

Appendix VI

Characteristic changes in the eye after death(domestic livestock)

Hours after death	First 30 mts after death	30 mts to 6 hrs 2hours)	6 hrs to 10 hrs (6hours)	11 hrs to 18 hrs (11 hours)
Diameter of pupil (in mm)	Fully dilated	15-16	11-12	6-7
Characteristics	1.Intraocular fluid and lens is transparent 2.Tapetum brilliant luminous green 3.Pupil size dependent on light response at the time of death, after a brief interval following death, the pupil is fully dilated	1.Slight loss of color and luminosity of the tapetum 2.Subtle wrinkling of cornea 3.Lens and fluid transparent	1.Lens and fluid develops opacity 2.Luminosity fades 3.Pupil narrows to one third or less of original diameter (use vertical diameter)	1. Opacity increases 2.Luminosity fades away 3. Pupil narrows to one-third or less of original diameter

After 30 hours : The color and pupil diameter remain the same.

Brown iris becomes a hazy blue after ~ 48 hours.

Freezing increases opacity, but halts pupil constriction. Thus, frozen carcasses may still be evaluated.

Appendix VII

DNA Studies

DNA studies have played a particularly important part in the battle against wildlife crime (Fain and Le May, 1995). They have contributed, for example, to linking a suspect with a crime scene, checking claims of captive-breeding, the identification of species and unequivocal gender determination.

General rules for sampling for DNA are as follows:

- Ascertain which available tissue may provide the most informative sample.

- Determine the purpose for which the tissue is required.

- Have a proper protocol and adhere to it.

- Remember that tissue samples from live animals are generally more reliable than from dead.

- Ensure that all instruments and containers are clean (sterile if possible) and free of 'stray' DNA.

- Wear gloves and handle tissues with forceps.

- Place tissues for DNA analysis in sterile plastic containers, paper bags or aluminium foil.

- Label samples correctly and indelibly with a unique number.

The method of transportation and storage of DNA samples is also important. The following method is recommended.

- Transport tissues preserved in ethanol or in bottles, etc., and as cold as possible preferably at 4ºC or –20ºC.

- Store frozen tissues at –70ºC if possible.

- Keep out of direct sunlight.

- In emergencies – keep cool, keep dry.

Although one of the advantages of PCR-based methods is that tissue samples that contain only limited amounts of DNA (or degraded DNA) can be examined, the value of any DNA study is greatly enhanced by using material of good quality. Quality will be reduced if there is contamination or if there is post-mortem change, when the DNA in cells begins to degenerate as a result of exposure to enzymes. The speed at which degradation occurs depends upon the type of tissue and the amount of moisture that is present in the cells – therefore DNA in liver will degrade rapidly compared with muscle on account of the presence of cellular enzymes. Blood shows very slow rates of degradation as long as the temperature is low. Sampling from carcasses and choice of tissue requires particular care. The ears of a dead animal can be particularly valuable as a source of DNA because they generally dry rapidly, thus DNA is preserved. The blood can be taken from all types of vertebrates. Other sources such as hair shafts (degrade slowly) are useful source of mitochondrial (mt) DNA.

DNA: is an acronym for deoxyribonucleic acid. In humans our DNA is comprised of ~3 billion DNA "building blocks" called A, T, C and G (referred to as "bases") which are strung together to form the human genome.

Genome: refers to the entire DNA sequence of A, T, C and G's in an organism (the human genome is ~3 billion bases in length divided around 23 pairs of chromosomes)

DNA sequencing: Is the process by which we determine the order of A, T, C and G's.

PCR: is an acronym for Polymerase Chain Reaction. It is a process to "photocopy" the DNA. It is critically important in forensics as it allows us to analyze very small starting quantities.

Mitochondrial DNA (mtDNA): Is a small ring of DNA (only 16,000 bases) that is always inherited from the mother and used for species identification purposes.

DNA tree: Is a way to visualize DNA sequences, rather than comparing similarities and difference by looking at the A, T, C and G code. It is common to present this graphically as a tree. The closer the things are on the tree the closer the relationships.

Nucleus: Is a compartment in a cell that contains most of the DNA (except mtDNA).

Microsatellites: Also called STR's. These are highly variable pieces of repeating DNA (e.g., ATATATATATATAT) scattered across your genome. When these are characterized they are often able to discriminate individuals.

DNA sexing: Males and females have a different genetic make up for some regions of their DNA, if these differences are characterized it is possible to determine if a sample is male or female. The chemical structure of everyone's DNA is the same. The only difference between people (or any animal) is the order of the *base pairs.* There are so many millions of base pairs in each person's DNA that every person has a different sequence. Using these sequences, every person could be identified solely by the sequence of their base pairs. However, because there are so many millions of base pairs, the task would be very time-consuming. Instead, scientists are able to use a shorter method, because of repeating patterns in DNA. These patterns do not, however, give an individual "fingerprint," but they are able to determine whether two DNA samples are from the same person, related people, or non-related people. Scientists use a small number of sequences of DNA that are known to vary among individuals a great deal, and analyze those to get a certain probability of a match.

DNA (deoxyribonucleic acid) represents the blueprint of the human genetic makeup. It exists in virtually every cell of the human body and differs in its sequence of nucleotides (molecules that make up DNA, also abbreviated by letters, A, T, G, C; or, adenine, thymine, guanine, and cytosine, respectively). The human genome is made up of 3 billion nucleotides, which are 99.9% identical from one person to the next. The 0.1% variation, therefore, can be used to distinguish one individual from another. It is this difference that can be used by forensic scientists to match specimens of *blood*, tissue, or hair follicles to an individual with a high level of certainty.

The complete DNA of each individual is unique, with the exception of identical twins. A DNA fingerprint, therefore, is a DNA pattern that has a unique sequence such that it can be distinguished from the DNA patterns of other individuals. DNA fingerprinting is also called DNA typing (.The two major uses for the information provided by DNA-fingerprinting analysis are for personal identification and for the determination of paternity).

DNA fingerprinting is based on DNA analyzed from regions in the genome that separate genes called introns. Introns are regions within a *gene* that are not part of the protein the gene encodes. They are spliced out during processing of the

Appendices

357

messenger RNA, which is an intermediate molecule that allows DNA to encode protein. This is in contrast to DNA analysis looking for disease causing mutations, where the majority of mutations involve regions in the genes that code for protein called exons. DNA fingerprinting usually involves introns because exons are much more conserved and therefore, have less variability in their sequence. In forensics laboratories, DNA can be analyzed from a variety of human samples including blood, semen, _saliva_, urine, hair, buccal (cheek cells), tissues, or bones. DNA can be extracted from these samples and analyzed in a laboratory and results from these studies are compared to DNA analyzed from known samples. DNA extracted from a sample obtained from a crime scene then can be compared and possibly matched with DNA extracted from the victim or suspect. DNA can be extracted from two different sources within the cell. DNA found in the nucleus of the cell, also called nuclear DNA (nDNA) is larger and contains all the information that makes us who we are. It is tightly wound into structures called chromosomes. DNA can also be found in an organelle within the cell called the mitochondria, which functions to produce energy that drives all the cellular processes necessary for life. Mitochondrial DNA (mtDNA) is much smaller, contains only 16,569 nucleotide bases (compared with nDNA, which contains 3.9 billion) and it is not wound up into chromosomes. Instead, it is circular and there are many copies of it. Nuclear DNA is analyzed in evidence containing blood, semen, saliva, body tissues, and hair follicles. DNA from the mitochondria, however, is usually analyzed in evidence containing hair fragments, bones, and teeth.

Mitochondrial DNA analysis is typically performed in cases where there is an insufficient amount of sample, the nDNA is uninformative, or if supplemental information is necessary. Unlike nDNA, where one copy of a _chromosome_ comes from the father and the other from the mother, mtDNA is exclusively inherited from the maternal side. Therefore, the maternal mtDNA should be the same as her offspring. This can be helpful in cases where it is not possible to obtain a sample from the suspect but it is possible to obtain a sample from one of the suspect's biologically related family members. By doing so, the suspect can be excluded as the culprit of a crime if the results indicate that the relevant family member's mtDNA does not match the mtDNA fingerprint from the sample.

DNA profiling techniques can provide information about a whole range of matters including

Parentage and kinship
Sex (gender)
Species
Population and strains.

DNA studies helps in linking a suspect with a crime scene, checking claims of captive breeding, the identification of species and unequivocal gender determination.

Practical Applications of DNA Fingerprinting

1. Paternity and Maternity - Because a person inherits his or her Variable number tandem repeat (VNTR)s from his or her parents, VNTR patterns can be used to establish paternity and maternity. The patterns are so specific that a parental VNTR pattern can be reconstructed even if only the children's VNTR patterns are known (the more the children produced, the more reliable the reconstruction). Parent-child VNTR pattern analysis has been used to solve standard father-identification cases as well as more complicated cases of confirming legal nationality and, in instances of adoption and biological parenthood.

2. Criminal Identification and Forensics - DNA isolated from blood, hair, skin cells, or other genetic evidence left at the scene of a crime can be compared, through VNTR patterns, with the DNA of a criminal suspect to determine guilt or innocence. VNTR patterns are also useful in establishing the identity of a homicide victim, either from DNA found as evidence or from the body itself.

3. Personal Identification - The notion of using DNA fingerprints as a sort of genetic bar code to identify individuals has been discussed, but this is not likely to happen anytime in the foreseeable future. The technology required to isolate, keep on file, and then analyze millions of very specified VNTR patterns is both expensive and impractical. Social security numbers, picture ID, and other more mundane methods are much more likely to remain the prevalent ways to establish personal identification.

Appendix VIII

Forensic Entomology (medical)

Is an area of Forensic Science that uses the insects or invertebrates found at a crime scene to provide information about the crime committed. Identification and analysis of insect specimens collected from an animal's carcass can allow the estimation of the minimum Post-mortem interval (PMI) as well as supply information on toxicological aspects of the body, wound –site location, interval of abuse or neglect or identification of victim and act as geographical trace evidence (whether the body has been moved after death). Eggs and fly larvae are found in and around the natural orifices of a body and on the underside when the carcass is lifted. They may also be collected from folds of skin behind the ears or in the crevices of joints. The soil underneath the carcass should be explored for any dispersing fly larvae, fly pupae or beetles to a depth of 3-5 cm.

It is the application and study of insect and other arthropod biology to criminal matters. Forensic entomology is primarily associated with Death investigations, used to detect drugs and poisons, Determine the location of an incident, Detect the length of a period of neglect in the elderly or children, and Find the presence and time of the infliction of wounds. Forensic entomology not only uses arthropod biology, but it pulls from other sciences introducing fields like chemistry and genetics, exploiting their inherent synergy through the *use of DNA in forensic entomology.*

Some types of invertebrates of possible significance in forensic work.

Arthropods: Some feed on the carcasses itself. Others on its inhabitants. Yet others use the remains of the animal or human as a home.

Insects: Coffin flies (Phoridae) are associated with different stages of decomposition.

Beetles: Carrion beetles are part of the faunal succession that responds to the presence of a dead animal. Some carrion (Silphidae) beetles are associated with late stages of decomposition or even skeletonization. Beetles, Skin beetles of Coleopterous species are associated with dry remains in later stages of decomposition or mummified carcasses.

Scarab dung beetles provide information of age, origin and sometimes health status of animal faeces. Beetles or their larvae die if faeces contain certain drugs. Butterflies and moths—various larvae may pupate on or adjacent to a carcass. Certain clothe moths may be feature of mummified bodies. Ants are often attracted at an early stage by blood and moisture. Fleas and lice may persist on carcass after death but move away as the temperature drops. If the carcass was poisoned or frozen, dead fleas and lice may be found. Millipedes are associated with late stages of decomposition, even skeletonization. Woodlice, crabs, shrimps, Daphnia etc are found on or in carcasses that are in water or damp places. Some may be inhaled.

Modern Techniques Used

1. Scanning electron microscopy : Usually fly larvae are used to aid in the determination of a PMI. However, sometimes the body may not contain maggots and only the eggs are present. In order for the data to be useful the eggs must be identified down to a species level to get an accurate estimate for the PMI. There are many techniques currently being developed to differentiate between. The various species of forensically important insects are identified based on by some of the morphological differences (presence/absence of anastomosis, the presence/absence of holes, and the shape and length of the median area). It may not be useful in a field study or to quickly identify a particular egg. The SEM method is good if ample time and resources to determine the species of the particular fly egg are available.

2. Potassium permanganate staining: A quicker and lower cost technique can be found in *potassium permanganate* staining. This process involves staining of collected eggs. They are rinsed with a *normal saline* solution and then moved to a glass petri dish. The eggs are then soaked in a 1% potassium permanganate solution for one minute. Then the eggs were dehydrated and mounted onto a slide for observation. These slides can be used with any *light microscope* with a calibrated eyepiece to compare various morphological features. The most important and useful features observed for identifying eggs are things like the size, length, and width of the plastron, as well as the morphology of the plastron in the area around the micropyle. The various measurements and observations are then compared to standards for forensically important species and used to determine the species of the egg.

3. Gene expression studies: Although physical characteristics and sizes at various *instars* have been used to estimate fly age, more recently a study has been conducted

Appendices

to determine the age of an egg based on the expression of particular genes. This is particularly useful in developmental stages that do not change in size, such as the egg or pupa, where only a general time interval can be estimated based on the duration of the particular developmental stage.

Essential Entomologist Equipment

- Nets
- Vials
- Fly traps
- Labels
- Pencils
- Camera
- Thermometer
- Forceps
- Ruler

Appendix IX

Laboratory-based investigations that may assist in forensic investigations

Discipline	Investigation	Information obtained
Parasitology	Identification of ecto and endoparasites, including blood parasites (haematozoa and microfilariae)	Species present may give clues as to geographical origin of host, time of death or contact with other species, including prey or predator
Haematology	Examination of blood smears Haemoglobin estimations	Presence of anucleate red cells will confirm that the blood came from a mammal. Other features may pinpoint the species more closely. Elevated values may indicate that an animal has originated from a high altitude
Histology	Standard methods	Many uses, e.g. anthracosis may indicate an industrial Origin, a giant cell focus a reaction to a foreign body

Appendix X

Life Spans of Mammals

Group	Common Name	Scientific Name	Years
Carnivores	Lion	Panthera leo	24
	Bobcat	Lynx rufus	34
	Tiger	Panthera tigris	20
	Leopard	Panthera pardus	20
	Jaguar	Panthera onca	20
	Puma	Felis concolor	16
	Fossa	Cryptoprocta ferox	17
	Coyote	Canis latrans	18
	Grey Wolf	Canis lupus	20
	Golden Jackal	Canis aureus	20
	Grey Fox	Vulpes cinereoargenteus	15
	Maned Wolf	Chrysocyon brachyurus	15
	Dhole, Asian Wild dog	Cuon alpinus	16
	Grizzly Bear	Ursus arctos	47
	Polar Bear	Ursus maritimus	21
Elephants	Asiatic or Indian Elephant	Elephas maximas	78
	African Elephant	Loxodonta africana	60
Antiodactyls	Dorcas Gazelle	Gazella dorcus	11.5
	Dama Gazelle	Gazella dama	15.5
	Llama	Lama glama	20
	Hippopotamus	Hippopotamus amphibius	42
	American Bison	Bison bison	22.5
	Defasser Waterbuck	Kobus ellipsiprymnus defassa	16.5
	Wild Boar	Sus scrofa	19.5
	Bactrian Camel	Camelus bactrianus	25.5
Perissodactyls	Burchill's Zebra	Equus burchelli	28
	Brazilian Tapir	Tapirus terrestris	30.5
	Indian Rhinoceros	Rhinocerus unicornis	47

Group	Common Name	Scientific Name	Years
Primates	Chimpanzee	Pan troglodytes	55+
	Orang-utan	Pongo pygmaeus	55+
	Gorilla	Gorilla gorilla	40+
	Rhesus Monkey	Macaca mulatta	29
	Yellow Baboon	Papio cynocphalus	45
Bats	Indian Flying Fox	Pteropus giganteus	30
	Vampire Bat	Desmodus rotundus	12
	Little Brown Bat	Myotis lucifugus	32
Rodents	Sumatran Crested Porcupine	Hystrix brachyura	27+
	African Porcupine	Hystrix cristata	20+
	Grey Squirrel	Sciuris carolinensis	14
	Malabar Squirrel	Ratufa indica	16
	Deer Mouse	Peromyscus maniculatus	5.5
	African Giant Rat	Cricetomys gambianus	4.5
	Chinchilla sp	Chinchilla lanigera	6.5

Gestation Days in Wild Animals

Animal	Gestation or incubation in days (average)	Animal	Gestation or incubation in days (average)
Ass	365	Kangaroo	32–39
Bear	180–240	Lion	105–113 (108)
Deer	197–300	Monkey	139–2701
Elephant	510 – 730(624)	Mouse	19–311
Fox	51– 63	Parakeet (Budgerigar)	17–20 (18)
Hippopotamus	220–255 (240)	Rat	21
Rabbit	30–35 (31)	Squirrel	44
Whale	365–5471	Wolf	60–63